P9-ARC-162

Foundations of
Quantum Chemistry

Foundations of
Quantum Chemistry

T. E. Peacock
University of Queensland
Australia

JOHN WILEY & SONS LTD
London   New York   Sydney   Toronto

Copyright © 1968 John Wiley & Sons Ltd.
All rights reserved. No part of this book may
be reproduced by any means, nor transmitted,
nor translated into a machine language without
the written permission of the publisher.

Library of Congress catalog card number 68–57665

SBN 471 67454 0

Printed in Great Britain by J. W. Arrowsmith Ltd., Bristol, England

Delaware Valley
College Library

541.28
P313

70 - 3883

Preface

This book contains the whole of a course in quantum chemistry as it developed over the years 1960 to 1966, during which time the author was a member of the staff of the Chemistry Department of King's College in the University of London. The course itself comprised about forty lectures, given at the rate of one a week through the second and first half of the third year course for the B.Sc. (Special) degree of the University of London. The book probably reflects the course as it was given for the last time in the session 1965–66. Since then the author has been engaged in teaching an entirely different course in the University of Queensland.

There has been no attempt to "blow up" the lecture notes into a lengthy text book four or five times the size of the present book. There are two reasons for this. First, it was the idea in writing the book to give what was essentially the course as delivered, secondly, the author hoped to produce a book which would be well within the limits of what an English undergraduate could afford to pay for a textbook for a single course of lectures.

There has been no attempt to reference the book in any manner. Names have been used where they are unavoidable, but extensive references do not belong to an undergraduate course. Readers will find a Bibliography at the end of the book which refers the reader to texts of varying degrees of sophistication on the various topics discussed in the book. These works are all well-referenced.

The author's own acknowledgments would make a list too long for enumeration, but he would however make acknowledgment to his many former colleagues at King's College, London, with whom much of the course was discussed. To single out one person is always difficult but special thanks must go to Professor D. W. G. Style, who gave the author time to teach the course as he wanted it taught, and who was a constant source of inspiration and a bottomless well of suggestions for its improvement. Then appreciation must go to the generations of undergraduates who endured the course and some of whom unwittingly made suggestions for its improvement.

Finally, thanks are due to the author's wife, Veronica, who has patiently endured many hours of loneliness and solitude whilst these pages were written and rewritten.

Queensland, 1968   T. E. PEACOCK

Contents

4 Small Molecules

5 Conjugated and Aromatic Molecules

6 Inorganic Molecules

1
The Principles of Quantum Mechanics

1.1 INTRODUCTION

During the course of the nineteenth century, the dynamics of bodies both large and small were assumed to obey Newton's laws of motion. In the case of the motion of charged particles, the classical electromagnetic equations together with Newtonian dynamics appeared adequate. With the turn of the century, however, the whole physical world-view changed.

In the case of very large bodies involved in motions through the heavens, Newton's laws were found to be inadequate. In the case of the very small particles which made up atoms and molecules the classical picture also failed. At both ends new theories were required. Einstein's relativity theories were found to be more successful in understanding stellar motion, and the early quantum theory of Planck was the beginning of the understanding of the behaviour of electrons.

As is well known, the first "successful" treatment of the electronic problem was carried out by Niels Bohr. In his treatment of the hydrogen atom, the electron was considered to move in a quantized orbit, whose motion was described by a combination of Newton's law and the Law of Electrostatic Attraction. The angular momentum was assumed to be quantized in units of $nh/2\pi$ where h is Planck's constant. This led to quantization of the energy and to the orbits being well-defined about the central nucleus. Whilst this theory gave a good explanation of the spectrum of the hydrogen atom and, with certain modifications, of the spectra of the alkali metals, application to atoms with more than one electron in the valence shell was a failure.

It rapidly became apparent that the Bohr theory could not produce quantitative results in these more complicated cases and for this reason the model was abandoned. Yet, if we consider the theory to be a failure we certainly misjudge it. The Bohr theory was necessary as the link between classical mechanics and quantum mechanics, and as such it had an important role to play.

The wave–particle duality of light was to provide the clue to a quantitative theory of electronic behaviour. Einstein's success in explaining the photoelectric effect in terms of light particles (photons) in no way invalidated the wave theory of light. The wave theory of light is just as essential to the explanation of interference and diffraction as the particle theory is necessary to explain the photoelectric effect.

In fact we never observe the light wave as such in any experiment. The waves are the mathematical expression of the behaviour of the photons. The motion of the photons is not described by Newton's laws but by the laws of wave motion. From this de Broglie saw that the wave–particle duality was a fundamental law of physics, but in anything much heavier than an electron the wave motion could be neglected.

Davisson and Germer showed that electrons can be diffracted and thus the wave theory of the electron was demonstrated. These experiments in 1927 confirmed de Broglie's hypothesis of particle–wave duality of the electron of 1924.

This approach led to the abandonment of the classical approach of the early quantum theory and to the development of a quantum mechanics almost independent of old classical concepts.

Schrödinger, in a series of papers published in 1926, used the equations of wave motion to describe the behaviour of the electron. All of the information about the electron is contained in the wave function ψ. The square of ψ (or more correctly $\psi^*\psi$, where ψ^* is the complex conjugate of ψ) expresses the probability of finding the electron at a given point in space. Since the dependence of the wave function upon time is known, a statistical knowledge can be obtained of the position of the electron at any time, but nothing more can be determined.

A different but entirely equivalent representation of quantum mechanics was formulated by Heisenberg, Jordan and others. This was based not on the wave equation but on the representation of observables by linear operators using a matrix representation. The observed values of these properties, e.g. angular momentum, energy, etc., are the eigenvalues of the matrices. This approach became known as "matrix mechanics" and von Neumann was able to demonstrate the equivalence between matrix and wave mechanics.

1.2 THE POSTULATES OF QUANTUM MECHANICS

The basic axioms of quantum mechanics are three in number. The first is concerned with the representation of observables by linear operators, the second with the observed values of the observable and the third with the mean of a series of measurements of the observable.

First Postulate: *To every observable there corresponds an operator.*

In particular, the operators of quantum mechanics are linear operators. It is important before going further to say what is meant by a linear operator. Let P be a linear operator and q the function on which P acts. Then we write Pq, to represent the action of P on q. If r is another function on which P acts, then the operator P is called a linear operator only if

$$P(q+r) = Pq + Pr \tag{1}$$

Not all operators are linear operators. We shall consider two examples, firstly the operator which tells us to "square" a function, i.e. ^{2}, and secondly the operator d/dx which tells us to differentiate a function. Let both operators act upon x^n. Then in the first case we have

$$P(x^n) = (x^n)^2 = x^{2n} \tag{2}$$

and in the second

$$P(x^n) = dx^n/dx = nx^{n-1} \tag{3}$$

Now let us consider the first case again where now, instead of having a single operand x^n, we have $x^n + x^q$.

$$P(x^n + x^q) = (x^n + x^q)^2 = x^{2n} + x^{2q} + 2x^{n+q} \tag{4}$$

According to the r.h.s. of (1) we have for this operation

$$P(x^n + x^q) = Px^n + Px^q = x^{2n} + x^{2q} \tag{5}$$

If the squaring operation was linear it would be expected that the result in (4) and (5) would be identical. Since clearly they are not, the operator ^{2} can not be a linear operator.

Let us now consider the second case again, and the effect of d/dx on $(x^n + x^q)$.

$$P(x^n + x^q) = d(x^n + x^q)/dx = nx^{n-1} + qx^{q-1} \tag{6}$$

Whilst

$$Px^n + Px^q = dx^n/dx + dx^q/dx = nx^{n-1} + qx^{q-1} \tag{7}$$

Since identical results are obtained in (6) and (7), the operator "to differentiate with respect to x" is clearly linear. In fact differential operators are extremely important in quantum mechanics.

Some of the more important operators in quantum mechanics are given in the table, for a particle j.

Observable	Classical representation	Quantum mechanical representation
Coordinate	x_j	x_j
Component of linear momentum	$p_{xj} = \dfrac{\mathrm{d}x_j}{\mathrm{d}t}$	$\dfrac{h}{2\pi i}\dfrac{\mathrm{d}}{\mathrm{d}x_j}$
Component of angular momentum	$m_j(y_j\dot{z}_j - z_j\dot{y}_j)$	$\dfrac{h}{2\pi i}\left(y_j\dfrac{\mathrm{d}}{\mathrm{d}z_j} - z\dfrac{\mathrm{d}}{\mathrm{d}y_j}\right)$
Energy	$\tfrac{1}{2}m_j(p_{xj}^2 + p_{yj}^2 + p_{zj}^2) + V_j$	$-\dfrac{h^2}{8\pi^2 m_j}\left(\dfrac{\partial^2}{\partial x_j^2} + \dfrac{\partial^2}{\partial y_j^2} + \dfrac{\partial^2}{\partial z_j^2}\right) + V_j$

The total energy of the n-particle system is then obtained by summing the energies of the individual particles, provided that there is no interaction between them.

If M_x, M_y and M_z represent the x, y and z components of angular momentum, then from the equation given in the last column of the table for the quantum-mechanical representation for M_x, it follows that

$$(M_x M_y - M_y M_x) = (-h/2\pi i)M_z$$
$$(M_y M_z - M_z M_y) = (-h/2\pi i)M_x \qquad (8)$$
$$(M_z M_x - M_x M_z) = (-h/2\pi i)M_y$$

In the table we have expressed the energy in its Hamiltonian form. The quantum mechanical operator derived from this is known as the Hamiltonian operator, and for a n-particle system it is defined by

$$H = -\frac{h^2}{8\pi^2}\sum_{j=1}^{n}\frac{1}{m_j}\cdot\left(\frac{\partial^2}{\partial x_j^2} + \frac{\partial^2}{\partial y_j^2} + \frac{\partial^2}{\partial z_j^2}\right) + V \qquad (9)$$

where V is the potential energy operator.

It is found that there is a relationship between the Hamiltonian and the angular momentum operators,

$$(HM_i - M_i H) = 0 \qquad (10)$$

H and the operators of angular momentum are said to "commute". This arises because H and M correspond to different observables. On the other hand the operators for the components of angular momentum (8) do not commute because they correspond to the same observable. This commutability of operators which are independent is an extremely important fact in quantum mechanics. We shall return to it at greater length later.

The exact form of the operator to represent any observable can only be found by trial and error, but of course most are now well known and have been well tested.

Second Postulate: *The only possible values which a measurement of an observable can have are given by the eigenvalues of the operator P, which represents the observable.*

Consider the operator P which may be represented by a matrix. The eigenvalues of the matrix P are the elements of the matrix after it has been diagonalized. In the case of the three-by-three matrix P in (11) we wish to find

$$\begin{pmatrix} \bar{p}_1 & 0 & 0 \\ 0 & \bar{p}_2 & 0 \\ 0 & 0 & \bar{p}_3 \end{pmatrix} \quad \text{from} \quad \begin{pmatrix} p_{11} & p_{12} & p_{13} \\ p_{21} & p_{22} & p_{23} \\ p_{31} & p_{32} & p_{33} \end{pmatrix} \tag{11}$$

where $\bar{P}$ the diagonal matrix is equivalent to the matrix P. That is P differs from $\bar{P}$ only by a unitary transformation. Let U be a matrix, and $U^\dagger$ be the complex conjugate of the transpose of U (the transpose of U is obtained by interchanging rows and columns of U). Now if $U^\dagger$ is the reciprocal of U, i.e. $U^\dagger U = UU^\dagger = 1$ (the unit matrix), then $U^\dagger PU = P'$. The two matrices are equivalent because they are connected by the unitary transformation, and this corresponds, for example, to a rotation in space or to a change in the basis functions defining the space (e.g. from Cartesians to spherical polars). The reverse transformation is also possible. If $U^\dagger PU = P'$, then pre-multiplying both sides by U and post-multiplying by $U^\dagger$, we get $UU^\dagger PUU^\dagger = UP'U^\dagger$, which simplifies to give $UP'U^\dagger = P$. One particular property of P and P', which is of fundamental importance, is that the sum of the diagonal elements (the "trace") of each matrix is the same, and we write

$$\sum_i P_{ii} = \text{tr } P = \sum_j P'_{jj} = \text{tr } \bar{P}' \tag{12}$$

This property is equivalent to the well known fact that under a rotation the length of a vector remains unchanged. We are particularly interested in the unitary transformation which converts the operator P into diagonal form $\bar{P}$.

Consider the set of simultaneous equations (13) derived from the elements of P:

$$(p_{11} - N)c_1 + p_{12}c_2 + p_{13}c_3 = 0$$

$$p_{21}c_1 + (p_{22} - N)c_2 + p_{23}c_3 = 0 \tag{13}$$

$$p_{31}c_1 + p_{32}c_2 + (p_{33} - N) = 0$$

where c_1, c_2 and c_3 are to be determined. The only non-trivial solution of these equations occurs when the determinant

$$\begin{vmatrix} (p_{11} - N) & p_{12} & p_{13} \\ p_{21} & (p_{22} - N) & p_{23} \\ p_{31} & p_{32} & (p_{33} - N) \end{vmatrix} = 0 \tag{14}$$

This leads to a polynomial in N of the third order. The roots of this equation are the eigenvalues of P, $\bar{p}_1$, $\bar{p}_2$ and $\bar{p}_3$. We substitute these values one at a time into the equations (13) and determine the ratios of c_1, c_2 and c_3 to each other. With the normalizing condition $c_{ij}^2 = 1$, the values of the coefficients are determined. Let c_{11}, c_{21} and c_{31} be the set corresponding to N_1, and similarly with the other two sets. Next arrange these coefficients into a matrix so that the columns are the coefficients of each particular N

$$\begin{pmatrix} c_{11} & c_{12} & c_{13} \\ c_{21} & c_{22} & c_{23} \\ c_{31} & c_{32} & c_{33} \end{pmatrix} = C \tag{15}$$

Then if P is a Hermitian matrix (i.e. $p_{ij} = p_{ji}^*$),

$$C^{\dagger}PC = \bar{P} \tag{16}$$

The columns (c_i) of C are known as the eigenvectors of P with the corresponding eigenvalues given by the appropriate elements of $\bar{P}$.

Example: To reduce a matrix to diagonal form, consider

$$\begin{pmatrix} 2 & 1 & 1 \\ 1 & 2 & 1 \\ 1 & 1 & 2 \end{pmatrix}$$

Form the set of simultaneous equations

$$(2 - N)c_1 + c_2 + c_3 = 0$$

$$c_1 + (2 - N)c_2 + c_3 = 0$$

$$c_1 + c_2 + (2 - N)c_3 = 0$$

Set up the determinant

$$\begin{vmatrix} (2-N) & 1 & 1 \\ 1 & (2-N) & 1 \\ 1 & 1 & (2-N) \end{vmatrix} = 0$$

Expanding the determinant we obtain

$$N^3 - 6N^2 + 9N - 4 = 0$$

which has roots 4, 1 and 1.

For $N = 4$, substituting in the equations we find

$$c_{11} = c_{21} = c_{31} = 1/\sqrt{3}$$

For $N = 1$, we obtain

$$c_{12} = -1/\sqrt{6} \qquad c_{22} = 2/\sqrt{6} \qquad c_{32} = -1/\sqrt{6}$$
$$c_{13} = 1/\sqrt{2} \qquad c_{23} = 0 \qquad c_{33} = -1/\sqrt{2}$$

The matrix C is

$$C = \begin{pmatrix} \dfrac{1}{\sqrt{3}} & -\dfrac{1}{\sqrt{6}} & \dfrac{1}{\sqrt{2}} \\ \dfrac{1}{\sqrt{3}} & \dfrac{2}{\sqrt{6}} & 0 \\ \dfrac{1}{\sqrt{3}} & -\dfrac{1}{\sqrt{6}} & -\dfrac{1}{\sqrt{2}} \end{pmatrix}$$

$$C^{\dagger}P = \begin{pmatrix} \dfrac{4}{\sqrt{3}} & \dfrac{4}{\sqrt{3}} & \dfrac{4}{\sqrt{3}} \\ -\dfrac{1}{\sqrt{6}} & \dfrac{2}{\sqrt{6}} & -\dfrac{1}{\sqrt{6}} \\ \dfrac{1}{\sqrt{2}} & 0 & -\dfrac{1}{\sqrt{2}} \end{pmatrix}$$

$$C^{\dagger}PC = \begin{pmatrix} 4 & 0 & 0 \\ 0 & 1 & 0 \\ 0 & 0 & 1 \end{pmatrix}$$

The eigenvalues p_i are the measured values of the observable represented by the operator P.

The operator P, the eigenvalues $\bar{p}_i$ and the eigenvectors c_i are related by the eigenvalue equation

$$Pc_i = \bar{p}_i c_i \tag{17}$$

This relation can be readily checked with the previously worked example.

The diagonalization of the matrix representing, say, the Hamiltonian for angular momentum is the crucial step in the matrix mechanics of Heisenberg and Jordan.

Differential equations whose solutions obey (17), in which the operator is of course a differential operator, are also known and form the basis of the Schrödinger wave mechanics with which we shall be concerned throughout the rest of this book.

Not all problems in quantum mechanics possess a discrete spectrum of eigenvalues; some cases are continuous. Special techniques have to be developed to deal with these cases.

In quantum mechanics it is usual to give the eigenfunction the symbol ψ. The function is required to satisfy two conditions

(a) $\int \psi^* \psi \, d\tau \leqslant \infty$ at all points, and
(b) ψ is single valued at all points in space.

The first condition is necessary to make the probability of finding the particle at any point in space finite. It is usual, as we have already assumed, to set this integral to unity, in other words we make the rational definition that the probability of finding the particle somewhere in space is unity. It is for similar reasons that ψ is required to be single-valued at all points in space.

Third Postulate: *When a given system is in a state ϕ, the expected mean of a sequence of measurements on the observable represented by the operator P, is given by*

$$\bar{p} = \int \phi^* p \phi \, d\tau \tag{18}$$

where we write $\bar{p}$ to be the mean value of a set of measurements $\bar{p}_i$.

This definition is defined as in statistics, where the expected mean of a set of measurements is given by

$$\bar{p} = \frac{1}{N} \sum_{i=j}^{N} p_i \tag{19}$$

The function ϕ in (18) is assumed to be normalized (i.e. $\int \phi^* \phi \, d\tau = 1$). It is important to notice that (18) does *not* predict the outcome of a single experiment.

The three postulates discussed above form the foundation of quantum mechanics.

In tackling a problem in wave mechanics it is first necessary to formulate the quantum-mechanical Hamiltonian. The wave equation

$$H\psi = E\psi \tag{20}$$

is then set up. Since this is a second order differential equation, solutions are not always easy to find and indeed in most cases do not exist in closed form. Very few problems have exact analytical solutions. The problems which can be solved in closed-form åre the "particle in a box", simple harmonic oscillator, rigid rotor and the hydrogen-like atom. For more complicated problems approximate methods have to be employed and we shall consider these methods later in this chapter. First, however, we shall deal with those cases where an exact solution is possible.

1.3 PROBLEMS WITH EXACT SOLUTIONS

1.3a The free particle

Consider a free particle moving in a constant potential. This potential may be taken as zero, since its sole effect to increase the energy by a constant amount V_0. Under these circumstances the Hamiltonian contains only the kinetic energy operator. The first step is to formulate this Hamiltonian. In classical mechanics the Hamiltonian is given by

$$H = \tfrac{1}{2}mv^2 = p^2/2m \tag{21}$$

The quantum mechanical Hamiltonian is obtained by replacing the momentum p by $h/2\pi i \times d/dx$ (assuming a one-dimensional model).

$$H = -\frac{h^2}{8\pi^2 m} \frac{d^2}{dx^2} \tag{22}$$

The wave equation is

$$-\frac{h^2}{8\pi^2 m} \frac{d^2\psi}{dx^2} = E\psi \tag{23}$$

Equation (23) can be integrated at once to give

$$\psi = N(E) \exp(ix\sqrt{2mE}/h) \tag{24}$$

The only restriction on the wave function (24) is that the wave function remains finite at all points in space. This is ensured if $\sqrt{2mE}$ is real and therefore E must always be positive. Apart from this restriction E is continuous and there is no quantization of energy in this case.

Using the third postulate discussed in the last section we can calculate the average value of the momentum of a free particle.

$$\bar{p} = \int \psi^* \left(-\frac{h^2}{8\pi^2 m} \frac{\partial^2}{\partial x^2} \right) \psi \, dx = \sqrt{2mE} = \sqrt{p^2} \tag{25}$$

It is equally acceptable to take as a solution of (23), the function

$$\psi = N(E) \exp\left(-\frac{i}{k} \sqrt{2mEx} \right) \tag{26}$$

in which case the average value of the momentum is $-\sqrt{p^2}$. The two solutions represent the particle moving in opposite directions.

In this example we have been concerned with stationary states, that is, the analogue of the standing wave in classical wave theory. Most of the problems in quantum mechanics of interest to the chemist are concerned with stationary states and we shall not discuss the time-dependent wave equation further at this point.

1.3b   The particle in a box

The next problem to consider is that of a freely-moving particle in a box of constant potential energy V, confined to the given volume by the presence of "walls" where the potential energy rises discontinuously to infinity. This is the simplest non-trivial problem in wave mechanics. The problem is shown diagrammatically in figure 1.

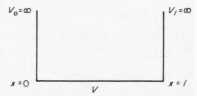

FIGURE 1. The particle in a box. The constant potential in the box is V rising to infinity at both sides.

Consider first the case where the particle is moving in a one-dimensional box of length l, i.e. $0 \leqslant x \leqslant l$. The constant potential energy can be set equal to zero. By the conditions of the problem the wave function is required to vanish at the box edges, i.e. when $x = 0$ or l. The Hamiltonian is

$$H = \frac{-h^2}{8\pi^2 m} \frac{d^2}{dx^2} \tag{27}$$

The wave equation is

$$\frac{-h^2}{8\pi^2 m} \frac{d^2\psi}{dx^2} = E\psi \tag{28}$$

or

$$\frac{d^2\psi}{dx^2} = -\frac{8\pi^2 m}{h^2} E \tag{29}$$

Writing p^2 for $8\pi^2 m/h^2$, the solution of (29) is

$$\psi = A \exp(ip\sqrt{E}x) + B \exp(-ip\sqrt{E}x) \tag{30}$$

An alternative way of writing (30) is

$$\psi = A' \cos p\sqrt{E}x + B' \sin p\sqrt{E}x \tag{31}$$

(using the relation $\exp(ix) = \cos x + i \sin x$).

At the point $x = 0$ we require ψ to vanish. Since cos 0 has the value unity this implies that A' vanishes. The other boundary condition requires ψ to vanish at the point $x = l$; this confines the argument of the sine term in (31) to integral multiples of $n\pi$, viz.

$$p\sqrt{E}l = n\pi \qquad (n = 1, 2, 3, \ldots) \tag{32}$$

or substituting for p we have

$$\frac{2\pi\sqrt{2m}}{h} \sqrt{E}l = n\pi$$

or

$$E = \frac{h^2 n^2}{8ml^2} \tag{33}$$

From (33) we see that the allowed values of E are discrete. In this case the energy is quantized. The eigenvalues remain discrete as $l \to \infty$, only becoming continuous when the length of the one-dimensional box is infinite, although long before this they have become very closely spaced. They are so closely spaced that we refer to the system of eigenvalues as quasi-continuous.

Substitution of (33) into (31) together with the requirement that A' has the value zero, gives

$$\psi = B' \sin n\pi x/l \tag{34}$$

To normalize the wave function we require that

$$\int_{-\infty}^{+\infty} \psi\psi \, dx = 1 \tag{35}$$

In this case we have

$$\int B'^2 \sin^2 n\pi x/l \, dx = B'^2 \int \sin^2 n\pi x/l \, dx = 1$$

Straightforward integration gives B' the value $\sqrt{2/l}$, and the wave functions (or eigenfunctions) of (28) are given by

$$\psi = \sqrt{\frac{2}{l}} \sin \frac{n\pi x}{l} \qquad (n = 1, 2, 3, \ldots) \tag{36}$$

From the form of the wave functions it can be seen that the wave function with $n = 1$ has no nodes, the function with $n = 2$ one node, with $n = 3$ three nodes and so forth. In other words the function with $n = k$ has $k - 1$ nodes.

It is now necessary to consider the case of the three-dimensional box, with sides of length l_x, l_y and l_z. We make the assumption that the solutions of the wave equation can be written in the form

$$\psi = X(x)Y(y)Z(z) \tag{37}$$

where X is a function of the variable x only, Y of y only and Z of z only. The importance of assuming this form of solution is that the wave equation factorizes into three equations: one in x, one in y and one in z. This method of solving differential equations in more than one variable is known as the separation of variables.

For the three-dimensional case, the Hamiltonian is

$$H = -\frac{h^2}{8\pi^2 m}\left(\frac{\partial^2}{\partial x^2} + \frac{\partial^2}{\partial y^2} + \frac{\partial^2}{\partial z^2}\right) + V_x + V_y + V_z \tag{38}$$

and the wave equation is

$$-\frac{h^2}{8\pi^2 m}\left(\frac{\partial^2}{\partial x^2} + \frac{\partial^2}{\partial y^2} + \frac{\partial^2}{\partial z^2}\right)\psi + (V_x + V_y + V_z)\psi = E\psi \tag{39}$$

Substituting (37) for ψ and dividing throughout (39) by (37) we obtain

$$\frac{1}{X(x)}\frac{\partial^2 X(x)}{\partial x^2} + \frac{1}{Y(y)}\frac{\partial^2 Y(y)}{\partial y^2} + \frac{1}{Z}\frac{\partial^2 Z(z)}{\partial z^2} + \frac{8\pi^2 m}{h^2}(E - V_x - V_y - V_z) = 0 \tag{40}$$

Rewriting (40) so that the l.h.s. is a function of x and y and the r.h.s. a function of z only, we have

$$\frac{1}{X(x)}\frac{\partial^2 X(x)}{\partial x^2} + \frac{1}{Y(y)}\frac{\partial^2 Y}{\partial y^2} + \frac{8\pi^2 m}{h^2}(E - V_x - V_y) = -\frac{1}{Z(z)}\frac{\partial^2 Z(z)}{\partial z^2} + \frac{8\pi^2 m}{h^2}V_z \tag{41}$$

Since the l.h.s. is a function of x and y only, if we vary x and y, keeping z constant, the r.h.s. of (41) is constant. Let us call this $(8\pi^2 m/h^2) \cdot E_z$. Then

$$\frac{\partial^2 Z(z)}{\partial z^2} + \frac{8\pi^2 m}{h^2}(E_z - V_z)Z(z) = 0 \tag{42}$$

leaving

$$\frac{1}{X(x)}\frac{\partial^2 X(x)}{\partial x^2} + \frac{1}{Y(y)}\frac{\partial^2 Y(y)}{\partial y^2} + \frac{8\pi^2 m}{h^2}(E - E_z - V_x - V_y) = 0 \tag{43}$$

Rewriting this so that the l.h.s. is a function of x only and the r.h.s. a function of y only, and following through the same reasoning, we obtain

$$\frac{\partial^2 Y(y)}{\partial y^2} + \frac{8\pi^2 m}{h^2}(E_y - V_y)Y(y) = 0 \tag{44}$$

and

$$\frac{\partial^2 X(x)}{\partial x^2} + \frac{8\pi^2 m}{h^2}(E_x - V_x)X(x) = 0 \tag{45}$$

where we have written $E = E_x + E_y + E_z$ in (45).

Equations (42), (44) and (45) are identical in form to the one-dimensional case if we assume $V_x = V_y = V_z$ is set to zero. The solutions can therefore be obtained without further analysis and are

$$E = \frac{h^2}{8m}\left(\frac{n_x^2}{l_x^2} + \frac{n_y^2}{l_y^2} + \frac{n_z^2}{l_z^2}\right) \tag{46}$$

and

$$= \sqrt{\frac{8}{l_x l_y l_z}}\sin\frac{n_x \pi x}{l_x}\sin\frac{n_y \pi y}{l_y}\sin\frac{n_z \pi z}{l_z} \tag{47}$$

This method of obtaining the solution is a common method of solving wave equations in two or more variables.

Of considerable interest is the particle moving in a "circular box", i.e. in the one-dimensional case where we bend the straight line of length l into a circle of circumference l. The solution of the problem differs from that of the one-dimensional box in that the cosine term in the wave function can no longer be disregarded, since the sole boundary condition is that of continuity around the circle, that is $\psi(x) = \psi(x + l)$. The solution of the problem is left as an exercise for the reader.

The solution of the "particle in a box" problem is of considerable significance, for it is on this model that the free-electron model of a metal is based. This model for the electrons in a metal was first put forward by Sommerfeld in the late 1920's. The solution of the problem is also of importance in the structure of molecules, where one model of conjugated and aromatic molecules is based on the solution of the "particle in a box" problem. This application will be discussed further in a later chapter.

1.3c The simple harmonic oscillator

Simple harmonic oscillation is a topic of considerable importance in quantum mechanics; it describes the motion of a particle which moves in a potential proportional to the square of its displacement. This gives rise to a restoring force proportional to the displacement.

The model of the simple harmonic oscillator is used to understand the vibrations of a diatomic molecule, to understand the vibrations of poly-atomic molecules and in the theory of solids to understand the vibrations in the solid and the evaluation of the electronic specific heat.

Consider first the one-dimensional oscillator. The classical Hamiltonian is

$$H = p^2/2m + kx^2/2 \tag{48}$$

The wave-mechanical Hamiltonian is

$$H = \frac{-h^2}{8\pi^2 m}\frac{d^2}{dx^2} + \tfrac{1}{2}kx^2 \tag{49}$$

and the wave equation is

$$\frac{d^2\psi}{dx^2} + \frac{8\pi^2 m}{h^2}(E - \tfrac{1}{2}kx^2)\psi = 0 \tag{50}$$

Making the substitutions $\alpha = 8\pi^2 mE/h^2$ and $\beta = 2\pi\sqrt{mk}/h$, (50) becomes

$$\frac{d^2\psi}{dx^2} + (\alpha - \beta^2 x^2)\psi = 0 \tag{51}$$

Further, if we make a variable $\xi = \sqrt{\beta}x$ then $d^2/dx^2 = \beta\, d^2/d\xi^2$, and (51) becomes

$$\frac{d^2\psi}{d\xi^2} + \left(\frac{\alpha}{\beta} - \xi^2\right)\psi = 0 \tag{52}$$

If we now write $\psi(\xi) = u(\xi)\exp(-\xi^2/2)$, (52) becomes

$$\frac{d^2 u}{d\xi^2} - 2\xi\frac{du}{d\xi} + \left(\frac{\alpha}{\beta} - 1\right)u = 0 \tag{53}$$

Finally, if we replace $(\alpha/\beta - 1)$ by $2n$, (53) then becomes one of the well-known differential equations. It is in fact Hermite's equation

$$\frac{d^2u}{d\xi^2} - 2\xi\frac{du}{d\xi} + 2nu = 0 \tag{54}$$

The requirement imposed on (53), in order that it becomes Hermite's equation, is that n be integral and therefore

$$\frac{\alpha}{\beta} = 2n+1 \tag{55}$$

from which we obtain the quantized energy levels of the harmonic oscillator since

$$\frac{\alpha}{\beta} = \frac{8\pi mE}{2h\sqrt{mk}} = \frac{4\pi E\sqrt{m}}{h\sqrt{k}}$$

and from (55) we have

$$\frac{4\pi E}{h}\sqrt{\frac{m}{k}} = 2n+1$$

from which

$$E = k\frac{h}{2\pi}\sqrt{\frac{k}{m}}(n+\tfrac{1}{2}) \qquad (n = 0, 1, 2, \ldots) \tag{56}$$

k is known as the force constant and nk/m is the vibrational quantum, where we have written $n = h/2\pi$. The state with $n = 0$, is the vibrational ground state. It can be seen from (56) that this state possesses energy $\tfrac{1}{2}hv$ (where we have set $v = 1/2\pi\sqrt{k/m}$, v is known as the vibrational frequency). This residual vibrational energy in the ground state is known as the zero-point energy. The existence of the zero-point energy can be considered as a consequence of the Uncertainty Principle, which states that one can never precisely know the position *and* momentum of a particle. The more accurately the one may be determined the greater is the uncertainty in the other. Heisenberg was led to formulate the Uncertainty Principle from considerations of the matrix mechanics solution for the simple harmonic oscillator.

The solutions of Hermite's equation (and hence the wave functions for the simple harmonic oscillator) are well known. These solutions of Hermite's equation, and hence of (54) are given by

$$H_n(\xi) = (-1)^n \exp(\xi^2)\frac{d^n\xi}{d\xi^2}\exp(-\xi^2) \tag{57}$$

The solutions of our equation are then

$$u_n(\xi) = H_n(\xi)$$

and the wave functions for the simple harmonic oscillators are

$$\psi_n = N^{-\frac{1}{2}} H_n(\xi) \exp\left(-\frac{\xi^2}{2}\right) \tag{58}$$

where $N^{-\frac{1}{2}}$ is the normalization constant. It is easy to see that (58) fulfills the requirement that the wave function disappears when x tends to infinity, for $\exp(-\xi^2/2)$ falls off more rapidly than the leading term of the Hermite polynomial H_n increases. The wave function then disappears as the variable tends to infinity. It can be shown that the normalization constant has the value $\sqrt{\beta/\pi/2^n n!}$, the normalized wave functions being

$$\psi = \left(\frac{\sqrt{\beta/\pi}}{2^n n!}\right) H_n(\xi) \exp(-\tfrac{1}{2}\xi^2) \tag{59}$$

$$(\xi = \sqrt{\beta} x)$$

The solution of the simple harmonic oscillator problem illustrates a common approach to the solution of wave equations. We seek to convert the wave equation into one of the standard differential equations of pure mathematics whose solutions are known. It is a technique which we use several times in the following sections.

The solution of the three-dimensional oscillator poses no new problems if we ignore the coupling between the oscillations in the directions of the basis vectors x, y and z. The three-dimensional oscillator solution is found by using the "separation of variables" method used in the last section and writing the solutions as a product of functions, each of which is a function of only one variable.

The anharmonic oscillator cannot be discussed within this framework because the wave equation in this case does not have an exact solution. It is necessary to use approximate methods for solving the wave equation for the anharmonic oscillator and we shall see later how this is done.

1.3d The rigid rotator

The theory of the rigid rotator is particularly important when we come to consider the rotation of a diatomic molecule in space. The problem to be solved is the rotation of two point masses m_1 and m_2 joined by a rigid, weightless link of length R, about the centre of gravity of the system. Let the distance of point 1 from the centre of gravity be r_1 and the distance of point 2 be r_2, then

$$m_1 r_1 = m_2 r_2$$

and

$$r_1 + r_2 = R$$

whence

$$r_1 = \frac{m_2 R}{m_1 + m_2}$$

and

$$r_2 = \frac{m_1 R}{m_1 + m_2}$$

To solve this problem it is most convenient to transform from Cartesian coordinates to spherical polars, by means of the transformation (see figure 2)

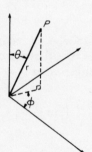

$$x = r \sin \theta \cos \phi$$
$$y = r \sin \theta \sin \phi$$
$$z = r \cos \theta$$

FIGURE 2.

r_1 and r_2 do not change during the rotation, so that the kinetic energy of the system is a function of θ and ϕ only. The kinetic energy of the first particle is given by

$$\tfrac{1}{2} m_1 \left[\left(\frac{\partial x_1}{\partial t} \right)^2 + \left(\frac{\partial y_1}{\partial t} \right)^2 + \left(\frac{\partial z_1}{\partial t} \right)^2 \right] = \tfrac{1}{2} m_1 r_1^2 \left[\left(\frac{\partial \theta}{\partial t} \right)^2 + \sin^2 \theta \left(\frac{\partial \phi}{\partial t} \right)^2 \right]$$

and the kinetic energy of the second particle m_2 is

$$\tfrac{1}{2} m_2 r_2^2 \left[\left(\frac{\partial \theta}{\partial t} \right)^2 + \sin^2 \theta \left(\frac{\partial \phi}{\partial t} \right)^2 \right]$$

The total kinetic energy of the rotating system is thus

$$T = \tfrac{1}{2}(m_1 r_1^2 + m_2 r_2^2)\left[\left(\frac{\partial \theta}{\partial t}\right)^2 + \sin^2 \theta \left(\frac{\partial \phi}{\partial t}\right)^2\right] \tag{60}$$

But $(m_1 r_1^2 + m_2 r_2^2)$ is simply the moment of inertia of the rotating system, usually given the symbol I, therefore

$$T = \tfrac{1}{2}I\left[\left(\frac{\partial \theta}{\partial t}\right)^2 + \sin^2\left(\frac{\partial \phi}{\partial t}\right)^2\right] \tag{61}$$

The quantum-mechanical operator can be written down bearing in mind that the kinetic energy operator must now be expressed in spherical polars. This can easily be obtained from any book dealing with transformations of coordinate systems* and it is found to be

$$T = -\frac{h^2}{8\pi^2 m}\left[\frac{1}{r^2}\frac{\partial}{\partial r}\left(r^2\frac{\partial}{\partial r}\right) + \frac{1}{r^2 \sin \theta}\frac{\partial}{\partial \theta}\left(\sin \theta \frac{\partial}{\partial \theta}\right) + \frac{1}{r^2 \sin^2 \theta}\frac{\partial^2}{\partial \phi^2}\right] \tag{62}$$

Since no external forces are acting on the system we can set $V = 0$, and since R is constant, the first term in (62) vanishes, and $r(= \tfrac{1}{2}R)$ can be regarded as unit length whence $m = I$, and the wave equation is

$$\frac{1}{\sin \theta}\frac{\partial}{\partial \theta}\left(\sin \theta \frac{\partial \phi}{\partial \theta}\right) + \frac{1}{\sin^2 \theta}\frac{\partial^2 \psi}{\partial \phi^2} + \frac{8\pi^2 IE}{h^2}\psi = 0 \tag{63}$$

This equation contains the two angular variables θ and ϕ. It is solved once again by the method of separation of variables. We look for a solution of the form

$$\psi = \Theta(\theta)\Phi(\phi) \tag{64}$$

Introducing this into (63) the wave equation becomes

$$\frac{\sin \Theta}{\Theta}\frac{\partial}{\partial \theta}\left(\sin \theta \frac{\partial \Theta}{\partial \theta}\right) + \frac{8\pi^2 IE}{h^2}\sin^2 \theta = -\frac{1}{\Phi}\frac{\partial^2 \Phi}{\partial \phi^2} \tag{65}$$

---

* See, for example, *The Mathematics of Physics and Chemistry*, Volume 1, Chapter 5, "Coordinate Systems, Vectors and Curvilinear Coordinates", by H. Margenau and G. M. Murphy, Van Nostrand Publishing Co., New York.

Setting both sides of (65) equal to a constant, say m^2, we obtain the pair of differential equations, each in one variable, given in (66)

$$\frac{d^2\Phi}{d\phi^2} = -m^2\Phi$$

$$\frac{1}{\sin\Theta}\frac{\partial\Theta}{\partial\theta}\left(\sin\theta\frac{\partial\Theta}{\partial\theta}\right) - \frac{m^2}{\sin^2\Theta}\Theta + \frac{8\pi^2 IE}{h^2}\Theta = 0$$

(66)

The first of these equations has the solution

$$\Phi(\phi) = c\exp(\pm im\phi) \tag{67}$$

This is an acceptable wave function provided only that m is an integer. This condition arises because Φ is required to be a single-valued function, which implies

$$\Phi(\phi) = \Phi(\phi + 2\pi)$$

or

$$\exp(im\phi) = \exp[im(\phi + 2\pi)]$$

This requires $\exp(2\pi mi)$ to be unity, which is only true if m is an integer. The normalization condition gives c the value $(2\pi)^{-\frac{1}{2}}$. The normalized solution of (67) is then

$$\Phi = \frac{1}{\sqrt{2\pi}}\exp(\pm im\phi) \qquad (m = 0, 1, 2, \ldots) \tag{68}$$

In the second equation of (66) we replace $8\pi^2 IE/h^2$ by $l(l+1)$. The equation then becomes

$$\frac{1}{\sin\theta}\frac{\partial}{\partial\theta}\left(\sin\theta\frac{\partial\Theta}{\partial\theta}\right) - \frac{m^2}{\sin^2\Theta}\Theta + l(l+1) = 0 \tag{69}$$

With a little rearrangement this equation takes on a well-known form, which has as its solutions the associated Legendre polynomials $P_l^{|m|}(\cos\theta)$, and which requires that l be integral and further that $l \geqslant |m|$. The normalized solutions are given by

$$\Theta(\theta) - \Theta_{l,\pm m}(\theta) = \sqrt{\frac{2l+1}{2}\frac{(l-|m|)!}{(l+|m|)!}}P_l^{|m|}(\cos\theta) \tag{70}$$

Legendre's equation has the form

$$(1-x^2)\frac{d^2u}{dx^2} - 2x\frac{du}{dx} + n(n+1)u = 0 \tag{71}$$

and has solutions

$$u = c\frac{d^n}{dx^n}(1-x^2)^n$$

The particular solution

$$u = P_n(x) = \frac{1}{2^n n}\frac{d^n}{dx^n}(x^2-1)^n \tag{72}$$

is known as the Legendre polynomial of degree n.

The associated Legendre equation is

$$(1-x^2)\frac{d^2u}{dx^2} - 2x\frac{du}{dx} + \left(n(n+1)-\frac{m^2}{1-x^2}\right)u = 0 \tag{73}$$

which has solutions

$$P_n^m(x) = (1-x^2)^{\frac{1}{2}m}\frac{d^m}{dx^m}P_n(x)$$

where the $P_n^m(x)$ are the associated Legendre polynomials, related to the Legendre polynomial P_n through the above equation. The associated Legendre polynomial $P_n^m(x)$ of degree n and order m is also given by

$$P_n^m(x) = \frac{(1-x^2)^{\frac{1}{2}m}}{2^n n!}\frac{d^{(n+m)}}{dx^{(n+m)}}(x^2-1)^n \tag{74}$$

To return to (69), the restriction on l leads to quantization of the energy, whose values are given by

$$E = \frac{h^2}{8\pi^2 I}l(l+1) \qquad (l = 0, 1, 2, \ldots) \tag{75}$$

The total wave function is then given by

$$\psi = \Theta_{l,\pm m}(\theta)\Phi_{\pm m}(\phi) = Y_{l,\pm m}(\theta, \phi) \tag{76}$$

The functions $Y_{l,\pm m}(\Theta, \phi)$ are called spherical harmonics. They are the eigenfunctions of the angular momentum operators M_z and M^2, being solutions of the eigenvalue equations

$$M^2 Y_{l,\pm m} = l(l+1)\frac{h^2}{4\pi^2}Y_{l,\pm m} \tag{77}$$

and

$$M_z Y_{l,\pm m} = \pm m\frac{h}{2\pi}Y_{l,\pm m} \tag{78}$$

We have seen therefore that the solution of the problem of the rigid rotator leads us directly to the solution of the angular momentum operators.

It is now necessary to discuss the simplification of the rigid rotator to the motion of the rotator in one plane, say the xy plane. In this case θ is constant, having the value 90°, so that the wave equation becomes

$$\frac{d^2\psi}{d\phi^2} + \frac{8\pi^2 IE}{h^2}\psi = 0 \tag{79}$$

The solutions of this equation are $\psi = \exp(\pm im\phi)$ and the eigenvalues are $\pm h^2 m^2/8\pi^2 I$. These eigenfunctions are also eigenfunctions of M_z having the eigenvalues $\pm nm$.

With one important exception we have now considered the important problems for which exact solutions can be obtained. The one exception is that of the hydrogen atom, which will be considered in the next chapter. Certain facts emerge from consideration of the eigenfunctions discussed in this section and it is now our concern to look further into these.

1.4 PROPERTIES OF EIGENFUNCTIONS

Apart from the fact that all of our eigenfunctions have been normalized to unity, there are certain other common properties. First, if we evaluate the integral

$$\int \psi_i^* \psi_j \, dv \tag{80}$$

it is found that in every case except where $i = j$, it has the value zero. We say that the eigenfunctions are orthogonal. It is a common property of a linear Hermitian operator that the eigenfunctions are orthogonal.

Secondly these eigenfunctions can be considered to be the basis functions (vectors) of a function (vector) space. The whole spectrum of eigenfunctions form what is called a complete set. To understand what is meant by the term "complete set", imagine an ordinary three-dimensional space. A point can be represented in this space by a vector with three components (the x, y and z components); such a vector is sometimes referred to as a 3-vector. Instead of choosing the three basis vectors as being in the x, y and z directions we could have made other choices, say r, θ and ϕ. However these functions are chosen, only three are required to define any point in the space. If we had only two, then it would be impossible to define a point in the space, if we had four then there would be too many, for any one can be expressed as a linear sum of the other three. The set of three vectors is called a complete set; the set of two vectors in this space is incomplete and the set of four vectors an overcomplete set. The spaces of quantum mechanics are much more complicated than these simple three-dimensional spaces, for they are spaces of n-dimensions and in many cases formally spaces of an infinite number of dimensions. Thus the eigenfunctions of an Hermitian operator form a complete set, and the first n-functions of an Hermitian operator can be considered to define a sub-space of the total space, and within the sub-space the n-eigenvectors form a complete sub-set.

It is also important to realize that any set of eigenvectors is not unique. Just as in a 3-space the Cartesian vectors can be transformed into (say) spherical polars by a unitary transformation or into any other set of three vectors, so the eigenvectors of the Hamiltonian can be transformed by a unitary transformation into another set, which are of course equivalent to the first set.

1.5   APPROXIMATE METHODS

Apart from a few simple cases, solution of the Schrödinger equation is impossible in closed form. Various approximate techniques have to be used in order that a solution may be obtained. There are two standard methods for obtaining approximate solutions and both depend on assuming the solution for a related simpler problem. For example, anharmonic terms can be taken into account as a small correction to the simple harmonic oscillator solution. The two methods of dealing with this problem are known as perturbation and variation theory.

1.5a   Perturbation theory

Suppose the solution of a problem, whose Hamiltonian is H is required. If there is a related situation, whose Hamiltonian is H^0 and whose solution is known, and H differs from H^0 by only a small amount, then the effect of $(H - H^0)$ can be taken into account by use of perturbation theory. The H^0 problem is then described as the zero-order Hamiltonian. Let H^0 have a spectrum of eigenvalues $E_0^0, E_1^0, \ldots, E_n^0$, with a set of eigenfunctions $\psi_1^0, \psi_2^0, \ldots$, then we can write

$$H^0 \psi_i^0 = E_i^0 \psi_i^0 \tag{81}$$

Since we have assumed that H is only slightly different from H^0, then H

can be written as a polynomial in a perturbation parameter λ.

$$H = H^0 + \lambda H' + \lambda^2 H'' + \lambda^3 H''' + \cdots \tag{82}$$

H' is known as the first-order perturbation correction, H'' as the second order perturbation and so on, the magnitude of the perturbation decreasing with increasing order. The eigenfunctions in (81) will of course form a complete orthogonal set.

If the perturbation is small then the true energy E_k of the kth level will lie very close to E_k^0, the value it has in the unperturbed case. We can therefore expand E_k as a power series in the perturbation parameter with E_k^0 as the first term.

$$E_k = E_k^0 + \lambda E_k' + \lambda^2 E_k'' + \cdots \tag{83}$$

Similarly, since the ψ_i^0's form a complete set, and the solution ψ_k must lie close to ψ_k^0, we can write the perturbed wave function for the kth level as a series in

$$\psi_k = \psi_k^0 + \lambda \psi_k' + \lambda^2 \psi_k'' + \cdots \tag{84}$$

If the perturbation is small (81), (83) and (84) will converge rapidly.

The wave equation for our perturbed system is

$$H\psi_k = E_k\psi_k \tag{85}$$

Substituting the series expansions for H, E and ψ we have

$$(H^0\psi_k^0 - E^0\psi_k^0) + (H^0\psi_k' + H'\psi_k^0 - E_k^0\psi_k' - E'\psi_k)\lambda$$
$$+ (H^0\psi_k'' + H'\psi_k' + H''\psi_k^0 - E_k^0\psi_k'' - E_k'\psi_k' - E_k'\psi_k' - E_k''\psi_k^0)\lambda^2 + \cdots \tag{86}$$

Here we have collected together coefficients of like powers of λ. Since λ is arbitrary, the coefficient of each power of λ must be zero. Equation (86) therefore simplifies into a set of equations, one for each power of λ. The first equation is of course the zero-order solution (81), the second equation gives the first-order perturbation solution, the third the second-order perturbation and so on. The zero-order, first-order and second-order equations are given in (87), (88) and (89).

$$H^0\psi_k^0 - E^0\psi_k^0 = 0 \tag{87}$$

$$H^0\psi_k' + H'\psi_k^0 - E_k^0\psi_k' - E_k'\psi_k = 0 \tag{88}$$

$$H^0\psi_k'' + H'\psi_k' + H''\psi_k^0 - E_k^0\psi_k'' - E_k'\psi_k' - E_k''\psi_k^0 = 0 \tag{89}$$

Let us examine the first-order equation (88). Assume that the first-order correction to the wave function can be written as a linear function of the wave functions of the unperturbed problem, viz,

$$\psi'_k = \sum_l a_l \psi^0_l \tag{90}$$

then

$$H^0 \psi'_k = \sum_l a_l H^0 \psi^0_l = \sum_l a_l E^0_l \psi^0_l \tag{91}$$

Putting (90) and (91) into (88) we obtain

$$\sum_l a_l (E^0_l - E^0_k)\psi^0_l = (E'_k - H')\psi^0_k \tag{92}$$

Multiplying each term by ψ^{0*}_k and integrating, then

$$\int \psi^{0*}_k \sum_l a_l (E^0_l - E^0_k)\psi^0_l \, dv = \int \psi^{0*}_k (E'_k - H')\psi^0_k \, dv \tag{93}$$

which leads to

$$\sum_l a_l (E^0_l - E^0_k) \int \psi^0_k \psi^0_l \, dv = E'_k \int \psi^{0*}_k \psi^0_k \, dv - \int \psi^{0*}_k H' \psi^0_k \, dv \tag{94}$$

Since the functions ψ^0_l are all orthogonal, we then obtain

$$0 = E'_k - \int \psi^{0*}_k H' \psi^0_k \, dv \tag{95}$$

which gives

$$E'_k = \int \psi^{0*}_k H' \psi^0_k \, dv \tag{96}$$

and writing H'_{kk} for $\int \psi^{0*}_k H' \psi^0_k \, dv$ we obtain

$$E_k = E^0 + H'_{kk} \tag{97}$$

It is now necessary to calculate the first-order correction to the wave function, and this implies determining the coefficients a_l in (90). To find a_l, multiply (92) by ψ^{0*}_l and integrate. Straightforward simplification and rearrangement of terms gives

$$a_l (E^0_l - E^0_k) = -\int \psi^{0*}_l H' \psi^0_k = -H'_{lk} \tag{98}$$

from which

$$a_l = -\frac{H'_{lk}}{E^0_l - E^0_k} \qquad (l \neq k) \tag{99}$$

This determines all the coefficients except a_k, which cannot be determined from (99). The perturbed wave function to first order is

$$\psi_k = N^{-\frac{1}{2}}\left(\psi_k^0 + \sum_l \frac{H'_{lk}}{E_k^0 - E_l^0}\psi_l^0\right) \tag{100}$$

This approach to determine the first-order energy and wave function is valid provided that none of the eigenvalues are degenerate, that is E_i^0 and E_l^0 do not have the same value for any pair of wave functions ψ_l^0 and ψ_i^0. If this is not the case, and degeneracies exist in the zero-order approximation, new techniques must be used to solve the perturbation problem.

In many cases the solution of the first-order perturbation equation is sufficient. The second-order correction is much smaller than the first-order correction and hence can usually be ignored with respect to the latter. The importance of it arises in cases where the first-order correction is zero.

The second-order perturbation correction to the energy and the wave functions are obtained by a straightforward method from (89) using the same methods as in the first-order case. In most cases it is the energy correction which is important rather than the wave functions, and this can be written down in terms of the zero-order and the first-order correction to the wave functions. In other words, by using (100) the second-order correction to the energy can be expressed as a function of the zero-order wave functions. The second-order correction E_k'' is found to be

$$E_k'' = \sum_{\substack{m \\ (\neq k)}} \frac{H'_{km}H'_{kn}}{E_k^0 - E_m^0} + H''_{kk} \tag{101}$$

Example: The Anharmonic Oscillator.
Consider the oscillator whose Hamiltonian is given by

$$H = -\frac{h^2}{8\pi^2 m}\frac{d^2}{dx^2} + \tfrac{1}{2}kx^2 + bx^4 \tag{102}$$

$$= H^0 + H'$$

where

$$H' = bx^4$$

The first three eigenfunctions of the harmonic oscillator are given in terms of the appropriate Hermite polynomials which are

$$H_0 = 1$$

$$H_1 = 2\xi$$

$$H_2 = 4\xi^2 - 2 \quad \text{where } \xi = \sqrt{\beta}x$$

These give as the first three normalized eigenfunctions

$$\psi_0 = \left(\sqrt{\frac{\beta}{\pi}}\right)^{\frac{1}{2}} \exp\left(-\frac{\xi^2}{2}\right)$$

$$\psi_1 = \left(\sqrt{\frac{\beta}{4\pi}}\right)^{\frac{1}{2}} 2\xi \exp\left(-\frac{\xi^2}{2}\right)$$

$$\psi_2 = \left(\sqrt{\frac{\beta}{64\pi}}\right)^{\frac{1}{2}} (4\xi^2 - 2)^{-\xi^2/2}$$

To find the effect of the perturbation bx^4 on the energy of the ground state, it is necessary to evaluate the integral $\int \psi_i H' \psi_j \, dv$.

$$H'_{00} = \sqrt{\frac{\beta}{\pi}} b \int_{-\infty}^{+\infty} \exp(-\xi^2) \frac{\xi^4}{\beta^2} \, dx$$

$$= \sqrt{\frac{1}{\pi}} b \int_{-\infty}^{+\infty} \exp(-\xi^2) \frac{\xi^4}{\beta^2} \, d\xi$$

$$= \frac{1}{\beta^2 \sqrt{\pi}} \int_{-\infty}^{+\infty} \exp(-\xi^2) \xi^4 \, d\xi$$

$$= \frac{2}{\beta^2 \sqrt{\pi}} \cdot \frac{3}{8} \sqrt{\pi}$$

$$= \frac{3b}{4\beta^2}$$

The energy of the anharmonic oscillator in the ground state is

$$E = E^0 + E'$$

$$= \tfrac{1}{2}h\nu + 3b/4\beta^2 \tag{103}$$

To calculate the first-order perturbed wave function, the integrals H'_{01}, and H'_{02} etc. must be evaluated. In a more precise calculation it would be necessary to include higher order terms (see equation 86). Simple integration leads to the values

$$H'_{01} = \frac{b}{\beta^2}\sqrt{\frac{2}{\pi}}, \quad \text{and} \quad H'_{02} = \frac{3b}{2\beta^2}$$

The perturbed wave function is

$$\psi_0 = N^{-1}\left(\psi_0^0 - \frac{b}{\beta^2}\sqrt{\frac{2}{\pi}}\frac{1}{h\nu}\psi_1^0 - \frac{3b}{2\beta^2}\frac{1}{2h\nu}\psi_2^0\right) \tag{104}$$

where

$$N^2 = \left(1 + \frac{2b^2}{\beta^4 \pi h^2 v^2} + \frac{9b^2}{16\beta^4 h^2 v^2}\right)$$

It is now necessary to consider the modifications which arise when the zero-order problem contains degenerate states. Suppose $\psi_i^0, \psi_j^0, \ldots, \psi_n^0$ are a set of eigenfunctions corresponding to the same eigenvalue E. Then we have

$$H\psi_i^0 = E\psi_i^0$$
$$H\psi_j^0 = E\psi_j^0 \qquad (105)$$

etc.

Furthermore any linear combination of these eigenfunctions is also an eigenfunction with eigenvalue E. So from the set of n degenerate eigenfunctions an infinite number of sets of n orthogonal functions can be constructed with the common eigenvalue E, viz.

$$\psi = \sum_{j=i}^{n} c_j \psi_j^0 \qquad (j = i, j, \ldots, n) \qquad (106)$$

One of the effects of the perturbation H' is to break the degeneracy of the n orthogonal functions. It is necessary to evaluate all the elements H'_{ij}. To find the perturbed energy levels and perturbed wave functions it is necessary to solve a set of n simultaneous equations

$$(H'_{ij} - E)c_j = 0 \qquad (i = i, j, \ldots, n) \qquad (107)$$

This set of equations has a non-trivial solution only if the determinant

$$\begin{vmatrix} (H'_{ij} - E) & H'_{ij} & \ldots & H'_{in} \\ \vdots & & & \\ H'_{ni} & & & (H'_{nn} - E) \end{vmatrix} \qquad (108)$$

vanishes. This determinant is known as a "secular determinant" and the set of equations (107) as "secular equations". The name arises because perturbation theory was first developed to calculate the effect of distant bodies on the orbits of planets and comets (Latin *saecula* = world). The roots E_i' of (108) and substitution of these one at a time into (107) allows the coefficients c_j to be determined. In fact this allows us to determine the ratios of the coefficients, but with the normalization condition the perturbed wave function can be determined, thus

$$\psi_k = \sum_{j=1}^{n} c_{jk}\psi_j^0 \tag{109}$$

where the c_{jk}'s are written in place of the unnormalized a_j's evaluated above.

The degeneracy will now be completely broken unless the perturbation possesses some or all of the symmetry of the zero-order problem; in which case some degeneracy may remain.

Perturbation theory is a powerful tool for obtaining approximate solutions to wave equations which have no exact solutions, provided that they differ from a known solution by only a small amount. Perturbations rapidly become too large for efficient convergence of the perturbation series and in these cases other methods must be sought.

1.5b Variation theory

The second method of obtaining approximate solutions to the wave equation is known as variation theory. Consider the integral $\int \phi^* H \phi \, dv$, where H is the Hamiltonian for the system and ϕ is some function which contains a variable parameter. We look for the turning points of this variation integral with respect to the parameter and these points give an upper limit to the energy of the system.

It is first necessary to show that the integral

$$E = \int \phi^* H \phi \, dv \tag{110}$$

is always an upper limit to the lowest energy level E_0 of the system.

The variation function ϕ is completely unrestricted and its choice may be arbitrary, but the more closely it is chosen to the physical reality which it represents, the more closely will E approach the true energy E_0. If by chance the variation function chosen happened to be the true ground state function, then

$$\int \phi^* H \phi \, dv = \int \psi_0^* H \psi_0 \, dv = E_0 \int \psi_0^* \psi_0 \, dv = E_0 \tag{111}$$

Expanding the variation function ϕ as a linear series in the complete set which forms the eigenfunctions of H, we have

$$\phi = \sum_i a_i \psi_i \tag{112}$$

where

$$\sum_n a_n^* a_n = 1 \quad \text{and} \quad a_m^* a_n \int \psi_m^* \psi_n \, dv = 0$$

Substituting (112) into (110) gives

$$E = \sum_{n,n'} a_n^* a_{n'} \int \psi_n^* H \psi_{n'} \, dv = \sum_n a_n^* a_n E_n \qquad (113)$$

since the ψ_i are the true eigenfunctions of H and

$$H\psi_n = E_n \psi_n \qquad (114)$$

Subtracting E_0 from both sides we obtain

$$E - E_0 = \sum_n a_n^* a_n (E_n - E_0) \qquad (115)$$

Since $E_n \geqslant E_0$ for all n, and $a_n^* a_n$ is always positive, the right hand side of (115) is always greater than zero, except in the one case where the variation function is the true ground-state wave function. Thus, we have

$$\sum_n a_n^* a_n (E_n - E_0) \geqslant 0$$

which implies $E - E_0 \geqslant 0$ and hence $E \geqslant E_0$, so that E is always an upper bound to the true ground state energy E_0.

At this stage we will not give examples of the application of the variation method since this method will be used continuously throughout the book.

One particularly important type of variation function is the linear variation function, which has the form

$$\phi = \sum_i c_i \chi_i \qquad (116)$$

in which the χ_i form a complete set (not necessarily orthogonal) known as basis functions. The coefficients c_i are the variation parameters and these must be varied to find the stationary value of the energy. This form of variation function is used to construct molecular orbitals from a linear sum of participating atomic orbitals (LCAO method, see Chapter 3).

We define the following symbols

$$H_{nn'} = \int \chi_n^* H \chi_{n'} \, dv \quad \text{and} \quad S_{nn'} = \int \chi_n^* \chi_{n'} \, dv \qquad (117)$$

$S_{nn'}$ is known as the overlap integral between χ_n and $\chi_{n'}$. When the basis set of functions are not orthogonal, the variation integral becomes

$$E = \frac{\int \phi^* H \phi \, dv}{\int \phi^* \phi \, dv} \tag{118}$$

Substituting (116) into (118) we obtain

$$E = \frac{\sum_{nn'} c_n^* c_{n'} H_{nn'}}{\sum_{nn'} c_n^* c_{n'} S_{nn'}} \tag{119}$$

which gives

$$E \sum_{nn'} c_n^* c_{n'} S_{nn'} = \sum_{nn'} c_n^* c_n H_{nn'} \tag{120}$$

We lose nothing by assuming that the coefficients c_n are real and we shall therefore assume this in what follows. To find the values of the coefficients c_i which minimize E, we differentiate E with respect to each c_i and set the resulting equation equal to zero, then if there are r basis functions χ_i we obtain a set of r simultaneous equations, each of the form

$$\frac{\partial E}{\partial c_k} \sum_{nn'} c_n c_{n'} S_{nn'} + E \frac{\partial}{\partial c_k} \sum_{nn'} c_n c_{n'} S_{nn'} = \frac{\partial}{\partial c_k} \sum_{nn'} c_n c_{n'} H_{nn'} \tag{121}$$

The condition for a minimum is that $\partial E / \partial c_k = 0$ for each k, and our set of simultaneous equations becomes

$$\sum_{n=1}^{r} c_n (H_{nk} - S_{nk} E) = 0 \qquad (k = 1, 2, \ldots, r) \tag{122}$$

The only non-trivial solution of this set of equations occurs when the determinant of coefficients $|(H_{nk} - S_{nk} E)|$ disappears. The values of the energy E_k can thus be determined by obtaining the roots of the polynomial in E which arises from expanding this determinant. Substituting the values of E_i into (122) the ratios of the c_n to each other can be obtained, and together with the normalization requirement

$$\sum_{n} c_n^2 + \sum_{nn'} 2 c_n c_{n'}^* S_{nn'} = 1$$

the corresponding wave function can be determined. The lowest eigenvalue is then the upper bound to the ground state and the corresponding wave function is the approximation to the ground-state wave function.

Provided that the Hamiltonian is not a function of the eigenfunctions, the eigenvalues $E_1, E_2, \ldots,$ are upper bounds to the first, second, ..., excited state energies of the system.

It is also possible to understand what happens to the energy levels when further basis functions are added to the variation function. This is shown in figure 3.

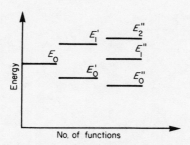

FIGURE 3. The effect of adding extra basis functions χ to the variation function.

Ideally further functions should be added until no further depression of the ground state energy occurs. At this point the best possible variation function attainable with the given type of basis function will have been obtained. In practice this is not often possible, but with the advent of high speed computers it is now becoming a feasible proposition. The linear variation function is a particularly good type of variation function, since it introduces into the wave function a flexibility which is difficult to obtain otherwise.

Other variation procedures exist which give lower bounds and both upper and lower bounds to the energy, but we shall not consider them here.

2

The Structure
of Atoms

The simplest atom is the hydrogen atom, in which a single electron moves in the field of a nucleus which has unit positive charge (in atomic units, where the charge on the electron is defined as the unit). The wave equation for the hydrogen atom can be solved in closed form. When the next atom in order of increasing complexity is considered, i.e. helium, which has two electrons moving in the field of the nucleus with charge two, the wave equation can no longer be solved in closed form and approximate methods, e.g. perturbation or variation theory, must be used to obtain a wave function.

2.1 THE HYDROGEN ATOM

Without loss of generality, the nucleus of the hydrogen atom can be considered to be at rest with respect to the much lighter and hence faster moving electron. The Hamiltonian for the system is

$$H = -\frac{h^2}{8\pi^2 m}\nabla^2 - \frac{e^2}{r} \tag{1}$$

where e is the charge on the electron (taken to be unity) and r is the distance of the electron from the nucleus. It is more correct to replace m, the mass of the electron, by the reduced mass $mM/m+M$ where M is the mass of the nucleus. Since $M = 1840m$, the reduced mass μ has the value $1840m/1841$ which is very close to m.

The wave equation is

$$\frac{h^2}{8\pi^2 m}\nabla^2\psi + \left(\frac{e^2}{r} + E\right)\psi = 0 \tag{2}$$

As in the case of the rigid rotor the problem is solved by expressing the wave equation in spherical polar coordinates rather than Cartesians. Further it is assumed that the solution has the form

$$\psi(r, \theta, \phi) = R(r)\Theta(\theta)\Phi(\phi) \tag{3}$$

in which case the wave equation separates into three equations each of one variable.

Following through the now familiar procedure, the following equations are obtained

$$\frac{d}{dr}\left(r^2\frac{dR}{dr}\right) + \frac{8\pi^2 m}{h^2}(Er^2 + e^2 r)R = l(l+1)R \tag{4}$$

$$\frac{1}{\sin\theta}\frac{d}{d\theta}\left(\sin\theta\frac{d\Theta}{d\theta}\right) + \left(l(l+1) - \frac{m^2}{\sin^2\theta}\right)\Theta = 0 \tag{5}$$

and

$$\frac{d^2\Phi}{d\phi^2} = -m^2\Phi \tag{6}$$

The solutions of (5) and (6) are the same as those found for the rigid rotor [(76), Chapter 1]:

$$\Theta(\theta)\Phi(\phi) = Y(\theta, \phi) \tag{7}$$

where Y is the spherical harmonic defined by the integers l and m.

After some manipulation (4) can be converted into a standard form. It is necessary to consider separately the cases where E is negative and positive. When E is negative (4) can be simplified by writing E in the form

$$E = -\frac{2\pi^2 me^4}{n^2 h^2} \tag{8}$$

and defining a new variable x related to r through

$$r = \frac{nh^2}{8\pi^2 me^2}x \tag{9}$$

Making these substitutions in (4) we obtain

$$\frac{d^2 R}{dx^2} + \frac{2}{x}\frac{dR}{dx} + \left[-\frac{1}{4} + \frac{n}{x} - \frac{l(l+1)}{x^2}\right]R = 0 \tag{10}$$

Looking for a solution of the form

$$R = u(x)x^l \exp(-\tfrac{1}{2}x) \tag{11}$$

it is found that $u(x)$ must satisfy

$$x\frac{d^2u}{dx^2}+(2l+2-x)\frac{du}{dx}+(n-l-1)u = 0 \tag{12}$$

This equation is identical with the differential equation whose solutions u are the associated Laguerre polynomials, if we set $\beta = 2l+1$ and $\alpha = n+l$.

$$x\frac{d^2u}{dx^2}+(\beta+1-x)\frac{du}{dx}+(\alpha-\beta)u = 0 \tag{13}$$

where $u = L_\alpha^\beta$ is the associated Laguerre polynomial and is related to L_α the Laguerre polynomial by

$$L_\alpha^\beta = \frac{d^\beta L_\alpha}{dx^\beta} \tag{14}$$

The Laguerre polynomial of degree α in x is defined by

$$L_\alpha(x) = e^x\frac{d^\alpha}{dx^\alpha}(x^\alpha e^{-x}) \tag{15}$$

Equation (13) possesses satisfactory solutions only if $(\alpha-\beta)$ is a positive integer. Since $\alpha-\beta = n-l-1$ and l may have the values $0, 1, 2, \ldots, n$ can have the values $1, 2, \ldots$, with the sole restriction that $n \geqslant l+1$. The allowed values of the energy E from (9) are given by

$$E = -\frac{2\pi^2 me^4}{n^2 h^2} \qquad (n = 1, 2, \ldots) \tag{16}$$

The r dependent part of the wave function is

$$R(r) = cx^l \exp(-\tfrac{1}{2}x)L_{n+l}^{2l+1}(x) \tag{17}$$

where

$$x = \frac{8\pi^2 me^2 r}{nh^2} \tag{18}$$

and c is the normalization factor.

The total wave function for the hydrogen atom is given by

$$\psi = R_{n,l}(r)Y_{l,\pm m}(\theta, \phi) \tag{19}$$

where the normalized spherical harmonic Y is given by

$$Y_{l,\pm m}(\theta, \phi) = \sqrt{\frac{2l+1}{4\pi}\frac{l-|m|!}{l+|m|!}}P^{|m|}(\cos\theta)\exp(\pm im\phi) \tag{20}$$

and the radial wave function $R(r)$ is by

$$R_{n,l}(r) = -\sqrt{\frac{(n-l-1)!}{2n[n+l!]^3}\left(\frac{2Z}{na_0}\right)^3}\left(\frac{2Zr}{na_0}\right)^l\exp\left(-\frac{Zr}{na_0}\right)L_{n+l}^{2l+1}\left(\frac{2Zr}{na_0}\right) \tag{21}$$

We have introduced here the radius of the first Bohr orbit in hydrogen as the unit of length, $a_0 = h^2/4\pi^2 m e^2$, and the nuclear charge Z which in the case of hydrogen is unity.

The spherical harmonics $Y_{l, \pm m}$ are eigenfunctions of the angular momentum operators M^2 and M_z. The eigenfunctions of H given by (19) are also eigenfunctions of M^2 and M_z. The functions (19) satisfy the eigenvalue equations

$$H\psi_{n,l,\pm m} = E_n \psi_{n,l,\pm m}$$

$$M^2 \psi_{n,l,\pm m} = l(l+1)\frac{h^2}{4\pi^2}\psi_{n,l,\pm m} \tag{22}$$

$$M_z \psi_{n,l,\pm m} = \pm m \frac{h}{2\pi}\psi_{n,l,\pm m}$$

Each eigenfunction is defined by the three quantum numbers n, l and m. The quantum number n determines the energy of the atom (as it does in the Bohr theory also), the quantum number l determines the total angular momentum of the atom and m determines the component of angular momentum along the z axis.

The normalized radial wave functions for the first three values of n are given in the table below. For convenience we have defined a new distance variable $\rho = (Z/a_0)r$, where Z is the nuclear charge, which for hydrogen is unity.

$n = 1,$	$l = 0$	$R_{1,0} = 2\left(\dfrac{Z}{a_0}\right)^{\frac{3}{2}} \exp(-\rho)$
$n = 2,$	$l = 0$	$R_{2,0} = \dfrac{1}{2\sqrt{2}}\left(\dfrac{Z}{a_0}\right)^{\frac{3}{2}} (2-\rho)\exp(-\rho/2)$
$n = 2,$	$l = 1$	$R_{2,1} = \dfrac{1}{2\sqrt{6}}\left(\dfrac{Z}{a_0}\right)^{\frac{3}{2}} \rho\exp(-\rho/2)$
$n = 3,$	$l = 0$	$R_{3,0} = \dfrac{2}{81\sqrt{3}}\left(\dfrac{Z}{a_0}\right)^{\frac{3}{2}} (27-18\rho+2\rho^2)\exp(-\rho/3)$
$n = 3,$	$l = 1$	$R_{3,1} = \dfrac{4}{81\sqrt{6}}\left(\dfrac{Z}{a_0}\right)^{\frac{3}{2}} (6\rho-\rho^2)\exp(-\rho/3)$
$n = 3,$	$l = 2$	$R_{3,2} = \dfrac{4}{81\sqrt{30}}\left(\dfrac{Z}{a_0}\right)^{\frac{3}{2}} \rho^2\exp(-\rho/3)$

From this table it can be seen that apart from the orbitals which have $l = 0$, all the orbitals have a node through the point $r = 0$, i.e. through the nucleus. From the two orbitals in the table which have $l = 1$ it can be seen that they have none and one node respectively for $n = 2$ and $n = 3$. In general an orbital has $n - l - 1$ nodes between the points $r = 0$ and $r = \infty$.

It will be remembered that the probability of finding an electron in a given orbital $\psi^m_{n,l}$ at a point lying between r and $r + dr$ from the nucleus, and having angular coordinates between θ and $\theta + d\theta$ and ϕ and $\phi + d\phi$, is

$$\psi^{m*}_{n,l}\psi^m_{n,l}\,d\tau = [R_{n,l}(r)]^2[Y_{l,\pm m}(\theta, \phi)]^2 r^2 \sin\theta\,dr\,d\theta\,d\phi \tag{23}$$

This probability can also be considered to be the density of the electron cloud at the point in space. These two alternative approaches to the meaning of $\psi^*\psi$ represent the particle–wave duality of the electron.

The total wave functions for the states $n = 2$, $l = 1$ are given below.

$n = 2,$	$l = 1,$	$m = 0$	$\psi_{2pz} = \dfrac{1}{4(2\pi)^{\frac{1}{2}}}\left(\dfrac{Z}{a_0}\right)\rho\exp(-\rho/2)\cos\theta$
$n = 2,$	$l = 1,$	$m = \pm 1$	$\psi_{2px} = \dfrac{1}{4(2\pi)^{\frac{1}{2}}}\left(\dfrac{Z}{a_0}\right)\rho\exp(-\rho/2)\sin\theta\cos\phi$
			$\psi_{2py} = \dfrac{1}{4(2\pi)^{\frac{1}{2}}}\left(\dfrac{Z}{a_0}\right)\rho\exp(-\rho/2)\sin\theta\sin\phi$

It can be seen that the wave functions as set out are strongly directed along the x, y and z axes. All three orbitals have the same energy, since energy is determined only by n when only one electron is present. Whilst these orbitals have a certain significance in understanding the structure of molecules it must be emphasized that there is no physical reason why these orbitals should be preferred to any other orthogonal linear combination of them.

Examine the electron density of the three orbitals with $l = 1$. From the above table we obtain

$$\sum_i \psi^x_{2pi}\psi_{2pi} = A^2_{2,1}R^2_{2,1}[\cos^2\theta + \sin^2\theta(\sin^2\phi + \cos^2\phi)]$$

$$= A^2_{2,1}R^2_{2,1} \qquad (i = x, y, z) \tag{24}$$

The electron density is determined only by the radial wave function and is spherically symmetrical in space. This is true for any set $n = n_i$, $l = l_j$. Such a set of orbitals with a given l is called a subshell and the total number of orbitals with a given n is known as a shell. It follows therefore that any atomic shell or subshell is spherically symmetric. The orbitals with $l = 0$ are known as s orbitals, with $l = 1$ as p orbitals and with $l = 2$ as d orbitals. The corresponding subshells are known as the s, p and d subshells respectively.

If one is interested in using the p orbitals for some problem in which the interest is in angular momentum properties rather than distribution of charge in space, it is found that the functions called $2p_x$ and $2p_y$ in the above table are not eigenfunctions of M_z. In fact the correct linear combinations are

$$\frac{1}{\sqrt{2}}(2p_x + i2p_y) \quad \text{and} \quad \frac{1}{\sqrt{2}}(2p_x - i2p_y)$$

which have eigenvalues of M_z of $+1$ and -1 respectively. Using these orbitals leaves the spatial charge distribution of the subshell unchanged, viz.

$$p^*_{+1}p_{+1} + p^*_{-1}p_{-1} + p_z^2 = \tfrac{1}{2}(p_x - ip_y)(p_x + ip_y) + \tfrac{1}{2}(p_x + ip_y)(p_x - ip_y) + p_z^2$$
$$= p_x^2 + p_y^2 + p_z^2$$

which as we have seen is spherically symmetrical.

So far we have been discussing the properties of the orbitals or eigenfunctions with negative eigenvalues or orbital energies. These are called bonding orbitals since an electron is bound to the nucleus to a greater or lesser extent in them.

By analogy to (8) we write the energies of the unbound states as

$$E = \frac{2\pi^2 me^4}{k^2 h^2} \tag{25}$$

and the radial component

$$r = \frac{kh^2 x}{8\pi^2 me^2}$$

The radial equation then becomes

$$\frac{d^2R}{dx^2} + \frac{2}{x}\frac{dR}{dx} + \left[\frac{1}{4} + \frac{k}{x} - \frac{l(l+1)}{x^2}\right]R = 0 \tag{26}$$

which at large values of x has the form

$$\frac{d^2R}{dx^2} + \frac{1}{4}R = 0 \tag{27}$$

which has solutions $R = c \exp(\pm \frac{1}{2}ix)$. These solutions have the necessary requirement of being finite at infinity and having a singularity at the origin, but solutions can be found which are finite at both bounds. The energy, however, is not quantized, all positive values being eigenvalues. The spectrum of eigenvalues is thus continuous and corresponds to the ionized state of the hydrogen atom.

The analysis given in the last few pages applies to all atoms with only one electron. It is for this reason that we have written the eigenfunctions in the above tables with the nuclear charge Z instead of unity, which is the value for the hydrogen atom. The orbitals given apply to He^+, Li^{2+}, Be^{3+} and so forth. In general the solution is known as the solution to the one-electron atom. With certain modifications the solutions can be used to discuss the cases of atoms which contain only one electron outside of a closed shell, e.g. Li, Be^+.

The ionization potential (the energy to remove the electron from the nucleus) is equal to the energy of the electron in the appropriate orbital. The ionization potential of the hydrogen atom in a $1s$ orbital is 13·6025 electron volts, for He^+ it is four times this and for Li^{2+} it is nine times the value for H, i.e. over 120 eV. It can be seen from this that the $1s$ orbital rapidly becomes a very tightly bound orbital. In more complex atoms the electrons in this orbital are so tightly bound that they can be factored out of the calculation with very little error.

The orbital energies found in the wave-mechanical theory of the hydrogen atom are just those found in the Bohr theory. It is also found that the contour of maximum electron density is coincident with the orbit of the Bohr calculation. This correspondence holds only for the one-electron case. For more complex atoms the Bohr theory fails as was discussed in Chapter 1.

2.2 THE HELIUM ATOM

The helium atom has two electrons moving in the field of a nucleus with charge two and now a new feature emerges, for not only is there interaction between the electrons and the nucleus, but there is an inter-action between the two electrons. It is this latter interaction which makes it impossible to solve the helium problem in closed form. If this interaction was non-existent or else small enough to be ignored, the problem could be solved by using the wave function in which each term in a simple product is a function of one-electron parameters only. This "one-electron approximation" is in fact used as the first approximation in the approach to the true solution. The interelectronic interaction is then included either through perturbation theory or else, as is more usual, by using variation theory and including a parameter, usually the nuclear charge, which is then optimized with respect to the energy in the usual way. The "Z" which results from this is generally known as the "effective nuclear charge" and is given the symbol Z'.

Assume then a variation function for the electronic wave function of the helium atom, which is a simple product of two $1s$ hydrogen-like orbitals in which the effective nuclear charge Z' is regarded as the variation parameter. The variation function is then

$$\phi = \phi_1 \phi_2 = \left(\frac{Z'^3}{a_0^3}\right) \exp(-Z'r_1/a_0) \exp(-Z'r_2/a_0) \tag{28}$$

in which Z' is the effective nuclear charge, r_1 and r_2 the distance of electrons 1 and 2 from the nucleus respectively. ϕ_1 is thus a function of electron 1 only and ϕ_2 of electron 2 only.

The Hamiltonian for the system is

$$h = -\frac{\hbar^2}{2m}(\nabla_1^2 + \nabla_2^2) - Ze^2\left(\frac{1}{r_1} + \frac{1}{r_2}\right) + \frac{1}{r_{12}} \tag{29}$$

Here we must use the correct nuclear charge Z to describe the interaction of the electrons with the nucleus.

Since ϕ_1 and ϕ_2 are hydrogen-like wave functions with nuclear charge $Z'e$, they are eigenfunctions of the corresponding hydrogen-like Hamiltonian h_1 or h_2, viz.

$$h_1\phi_1 = \left(-\frac{\hbar^2}{2m}\nabla_1^2 - \frac{Z'e^2}{r_1}\right)\phi_1 = -Z'^2 E_H \phi_1 \tag{30}$$

where $-E_H$ is the energy of the $1s$ orbital of hydrogen. A similar equation holds for ϕ_2. From (30) we can substitute for the kinetic energy operator in (29) and obtain for the variation integral $E = \int \phi h \phi \, d\tau$

$$E = -2Z'E_H + (Z' - Z)e^2 \int_0^\infty \phi^* \left(\frac{1}{r_1} + \frac{1}{r_2}\right) \phi \, d\tau + \int_0^\infty \frac{\phi^*\phi}{r_{12}} \, d\tau \tag{31}$$

The first integral in (31) becomes upon substitution of (28)

$$e^2 \int_0^\infty \phi^*\left(\frac{1}{r_1} + \frac{1}{r_2}\right)\phi \, d\tau = e^2 \int_0^\infty \left(\frac{\phi_1^2}{r_1} + \frac{\phi_2^2}{r_2}\right) d\tau = 2e^2 \int_0^\infty \frac{\phi_1^2}{r_1} \, d\tau_1 \tag{32}$$

The last integral in (32) is easily evaluated and gives $Z'e^2/a_0$ or $2Z'E_H$, the total contribution being $4Z'E_H$. The second integral in (31) is not so easy to evaluate and involves an expansion of r_{12}. It is found to have the value $\frac{5}{4}Z'E_H$.

Inserting these values of the integrals into (31) the variation integral has the value

$$E = \{-2Z'^2 + \tfrac{5}{4}Z' + 4Z'(Z'-Z)\}E_H \tag{33}$$

The value of Z' which optimizes the energy is found by differentiating E with respect to Z' and setting $dE/dZ' = 0$. This gives

$$(-4Z' + \tfrac{5}{4} + 8Z' - 4Z)E_H = 0 \tag{34}$$

from which we obtain

$$Z' = Z - \tfrac{5}{16} \tag{35}$$

Substitution of this value for the effective nuclear charge into (31) gives an energy of

$$E = -2(Z - \tfrac{5}{16})^2 E_H$$

This energy is much better than the value one obtains using first-order perturbation theory. The quantity $\frac{5}{16}$ in (35) is known as the screening constant and is usually designated by the symbol σ. It represents the screening of the nucleus by the other electron. The nucleus appears to an electron to have a charge of only $1\frac{11}{16}$ instead of the true charge of two.

The $1s$ hydrogen-like orbitals with Z' of $1\frac{11}{16}$ are known as one-electron orbitals. To choose a variation function of this type is a considerable advantage since the function retains a physical reality which would be lost by using a variation function of another kind. It is possible to construct a variation function out of any set of functions provided only that the set is complete. Whilst such sets give variation functions which give a more accurate energy than hydrogen-like functions they have little physical interest, though with the development of large electronic digital computers, current interest is turning away from the functions which have considerable physical significance to functions which can be handled quicker and more easily by these machines.

2.3 ELECTRON SPIN

If spectroscopic data is used to construct the energy level system for an alkali metal, which has one electron outside a closed shell, it is found that the levels appear to occur in very closely-spaced groups of two. In the alkaline earth series where there are two electrons outside a closed shell it is found that the levels appear to be in two different sets, one with single levels and the other with groups of three closely spaced lines. If the electron possesses only orbital angular momentum and a principal quantum number n, which alone determines the energy, then no explanation for this multiplicity of lines can be found.

An explanation of this occurrence was given by Goudsmit and Uhlenbeck in 1925 in the framework of the old Bohr-type quantum theory. This theory postulated the existence of an intrinsic angular momentum associated with the electron apart from the angular momentum which arises from the motion of the electron about the nucleus. This intrinsic angular momentum possesses a magnitude of $h/2$ and has associated with it a magnetic moment of magnitude $eh/4\pi mc$. Both may be ascribed to the rotation of the electron about its own axis, and this gave rise to the concept of the spinning electron. The interaction of the magnetic moment of the spinning electron with the magnetic field produced by the motion of the electrons in their orbits is responsible for the observed pattern of levels. In the case of the alkali metals, the angular momentum of the single electron can be aligned either with or against the orbital angular momentum; in the case of the alkaline earth metals the spin of the electrons can either be in the same or opposite direction. From this picture emerges the possibility of the levels lying either as singlets (singly spaced lines) or triplets (sets of three lines closely spaced).

The existence of electron spin was first demonstrated in an experiment by Stern and Gerlach in 1922. A beam of silver atoms was passed through a strong non-uniform magnetic field and was split into two distinct beams. Had there been no spin associated with electrons, the beam would have broadened but would not have split in this way. However, if electron-spin angular momentum existed and was quantized, splitting would have occurred (into as many beams as were consistent with the quantization). The splitting of the beam into two demonstrated that only two values for the spin angular momentum are possible. These values are associated with an angular momentum whose z component takes the values $\pm\frac{1}{2}$ angular momentum units ($\hbar$).

From this it is expected that the spin angular momentum has associated with it a set of angular momentum operators analogous to the set M_x, M_y, M_z and M^2. This set of operators has the symbols S_x, S_y, S_z and S^2. There is only one eigenvalue of S^2 namely $(1+\frac{1}{2})\frac{1}{2}\hbar^2$ and two eigenvalues of $S_z \pm \frac{1}{2}\hbar$. The eigenfunction corresponding to $+\frac{1}{2}\hbar$ is given the symbol α and the eigenfunction corresponding to $-\frac{1}{2}$ the symbol β. The spin eigenfunctions satisfy the following eigenvalue equations:

$$S^2\alpha = \tfrac{3}{4}\hbar^2\alpha \qquad S^2\beta = \tfrac{3}{4}\hbar^2\beta$$
$$S_z\alpha = \tfrac{1}{2}\hbar\alpha \qquad S_z\beta = -\tfrac{1}{2}\hbar\beta$$

$$(36)$$

The spin operators obey the same commutation laws as were given in Chapter 1 for angular momentum, viz.

$$(S_x S_y - S_y S_x) = i S_z \tag{37}$$

and

$$(S^2 S_z - S_z S^2) = 0 \tag{38}$$

The spin eigenfunctions form an orthonormal set:

$$\int_{-\infty}^{\infty} \alpha^*(1)\beta(1)\,d\tau_1 = \int_{-\infty}^{\infty} \beta^*(1)\alpha(1)\,d\tau_1 = 0$$

$$\int_{-\infty}^{\infty} \alpha^*(1)\alpha(1)\,d\tau_1 = 1, \qquad \int_{-\infty}^{\infty} \beta^*(1)\beta(1)\,d\tau_1 = 1 \tag{39}$$

(The number in brackets refers to the electron in question and the integration is over the space of that electron.)

The state of an electron is therefore defined by four quantum numbers, n, l, m and s and not by the three n, l and m which arise in the solution to the hydrogen atom problem.

It was not necessary to consider electron spin explicitly in the discussion of hydrogen and helium, because the Hamiltonian is "spin free", that is, the total energy of the atom is considered to be independent of the electron spin. Whilst this is a good approximation for the lightest atoms, the interaction between the spin and orbital angular momentum rapidly becomes important as heavier and heavier atoms are considered and a term to represent this interaction has then to be included.

2.4 THE PAULI PRINCIPLE

Before the advent of quantum mechanics, Pauli enunciated the general principle that no two electrons in the same atom could have the same set of four quantum numbers n, l, m and s. For example, if two electrons have the same values of n, l and m, that is they occupy the same orbital, then they must have different values of the spin component operator S_z. A more complete wave function for the ground state of helium, which has both electrons contained in the $1s$ orbital, is

$$1s(1)\alpha(1)1s(2)\beta(2) \tag{40}$$

Since electrons are indistinguishable, an equally acceptable wave function is

$$1s(1)\beta(1)1s(2)\alpha(2) \tag{41}$$

A better wave function than either (40) or (41) will be one in which both are included, viz.

$$\frac{1}{\sqrt{2}}(1s(1)\alpha(1)1s(2)\beta(2) \pm 1s(1)\beta(1)1s(2)\beta(2)) \tag{42}$$

It is found that electrons, in common with all particles with half-integral spin (collectively called fermions), must have wave functions which change sign when two electrons are interchanged. The correct wave function for the ground state of the helium atom is

$$\frac{1}{\sqrt{2}}(1s(1)\alpha(1)1s(2)\beta(2) - 1s(1)\beta(1)1s(2)\alpha(2)) \tag{43}$$

Since (40) and (41) both satisfy the original formulation of the Pauli principle then clearly it must be reformulated to include the fact that electronic wave functions are antisymmetric with respect to interchange of electrons. A more satisfactory formulation is: "The wave function representing an actual state of an electronic system must be completely antisymmetric in the coordinates of any two electrons; i.e. on interchanging the coordinates of any two electrons the wave function must change sign."

It is more conventional to write an electronic wave function as a product of two functions: one involving the space coordinates and the second the spin coordinates only. Equation (43) when rewritten in this way becomes

$$1s(1)1s(2)\frac{1}{\sqrt{2}}(\alpha(1)\beta(2) - \beta(1)\alpha(2)) \tag{44}$$

2.5 EXCITED STATES OF HELIUM

Consider the wave function corresponding to the state of the helium atom in which one electron has been excited from the doubly occupied $1s$ orbital to the $2p$ orbital. Without including electron spin this wave function is

$$[1s(1)2p(2) \pm 2p(1)1s(2)] \tag{45}$$

The two electrons are no longer constrained to have opposing spins. The possible spin eigenfunctions for systems of two electrons are

$$\text{(i) } \frac{1}{\sqrt{2}}[\alpha(1)\beta(2) - \beta(1)\alpha(1)] \qquad \text{(iii) } \alpha(1)\alpha(2)$$

$$\text{(ii) } \frac{1}{\sqrt{2}}[\alpha(1)\beta(2) + \beta(1)\alpha(2)] \qquad \text{(iv) } \beta(1)\beta(2)$$

(46)

The first spin function is an eigenfunction of S^2 with the eigenvalue 0, the last three are eigenfunctions of S^2 with the eigenvalue $2\hbar^2$ (i.e. $1(1+1)\hbar^2$). The first two are eigenfunctions of S_z with eigenvalue 0 and the last two eigenfunctions of S_z with eigenvalues $+1\hbar$ and $-1\hbar$. The first is a singlet state and corresponds to a spin eigenfunction which is antisymmetric with respect to interchange of the electrons. The last three are the three components of a triplet with $S = 1$ (the spin quantum number analogous to l) and components $S_z = 1, 0$ and -1.

In order that the total wave function be antisymmetric it is necessary that the singlet spin function be combined with the plus sign in (45), and the triplet spin function (the components of which are all symmetric with respect to interchange) with the antisymmetric space function in (45). The total wave functions for state of helium $(1s)(2p)$ are given in equations (47) and (48).

Singlet

$$\frac{1}{\sqrt{2}}[1s(1)2p(2) + 2p(1)1s(2)]\frac{1}{\sqrt{2}}[\alpha(1)\beta(2) - \beta(1)\alpha(2)]$$

(47)

Triplet

$$\frac{1}{\sqrt{2}}[1s(1)2p(2) - 2p(1)1s(2)] \begin{cases} \alpha(1)\alpha(2) \\ \frac{1}{\sqrt{2}}[\alpha(1)\beta(2) + \beta(1)\alpha(2)] \\ \beta(1)\beta(2) \end{cases}$$

(48)

2.6 THE MANY-ELECTRON ATOM

When there are more than two electrons, the calculation of the energy of the ground state (and of other states) follows on the lines discussed in the previous sections. However there is a complicating factor, for the interelectron repulsions become much more complicated to enumerate. The writing down of a suitable antisymmetric wave function presents at first sight a very difficult problem.

The electronic structure of many-electron atoms can be understood in terms of hydrogen-like orbitals. As an initial approximation, it is assumed that two electrons with opposite spins can be fed into each orbital starting with the one of lowest energy (1s) and continuing in order of increasing energy. This method of building up the structure of atoms is known as the Aufbau ("building up") principle. When two or more electrons are contained in the atom the p orbitals are no longer degenerate with the s orbital of the same quantum number n. Degeneracy occurs when two or more orbitals have the same energy. Similarly each group of orbitals with a given n, l acquires a different energy compared with the groups of orbitals with the same n but different l. The energy order is $1s < 2s < 2p < 3s < 3p < 4s < 3d$. When one arrives at the heavier atoms the energy order is often changed, particularly among p, d and f electrons.

The effective nuclear charge Z' can be written down very easily by using a set of rules known as "Slater's rules", after J. C. Slater who first formulated them. The shielding constant σ for a given orbital is determined in the following way. The electrons are divided into the usual groups $1s$, $2s$, $2p$, $3s$, $3p$, $3d$, $4s$, $4p$, $4d$, $4f$, ... The shielding constant for an electron in a given orbital is made up from the following contributions.

   (a) Zero from any shell outside the one considered.
   (b) 0·35 from every other electron in the shell considered (except $1s$ where the value is 0·30).
   (c) If the subshell considered is s or p, then 0·85 from each electron with a principal quantum number less by one, and 1·00 from every electron further in. If the electron is a d or f electron then 1·00 from each electron further in.

For a $2p$ electron in carbon, the shielding constant is 2×0.85 (for the $1s$) and 3×0.35. This gives a total shielding constant of 2·75 and an effective nuclear charge of 3·25. For a $1s$ electron the effective nuclear charge is 5·70.

A further rule for quantum shells with principal quantum number n of four or over defines values for the effective principal quantum number n^*. They are

$$n = \quad 4 \quad\quad 5 \quad\quad 6$$

$$n^* = 3.7 \quad 4.0 \quad 4.2$$

The n^* take into account the shielding and interaction of electrons amongst themselves.

Use of Slater's rules enables us to write down fairly reliable exponents for hydrogen-like orbitals. More sophisticated methods exist for writing down orbitals where greater accuracy is required.

Having written down an acceptable set of orbitals for the atom concerned, the next step is to write an acceptable wave function from them. In the first instance, consider the case where the configuration concerned has either a closed shell (the noble gases and certain ions) or at least a closed subshell (say six p electrons, ten d electrons, etc.). The easiest way to write down an antisymmetric function is to express it as a determinant if this is possible. When a determinant is used to represent an electronic wave function it is known as a "Slater determinant", again after Slater, who was the first to use these wave functions. A Slater determinant is constructed in the following way. Consider the case of the fluorine ion F^- where we have the electronic configuration $(1s)^2(2s)^2(2p_x)^2(2p_y)^2(2p_z)^2$, a closed shell configuration with ten electrons. Let the symbols $1s$ etc. represent the appropriate wave functions. The Slater determinant is then as shown on page 47.

The determinantal wave function is completely antisymmetrical, because interchange of two electrons implies interchange of two columns of the determinant which changes the sign of the determinant, and hence the sign of the wave function.

A more concise way of writing a determinantal wave function is to write down only the leading or diagonal term, which in the above case is

$$\frac{1}{\sqrt{N!}}|1s(1)\alpha(1)1s(2)\beta(2)2s(3)\alpha(3)2s(4)\beta(4)2p_x(5)\alpha(5)2p_x(6)\beta(6)$$

$$2p_y(7)\alpha(7)2p_y(8)\beta(8)2p_z(9)\alpha(9)2p_z(10)\beta(10)| \qquad (49)$$

Often the elements of the determinant are abbreviated further by using the notation $1s(N)$ to denote the electron in the $1s$ orbital with α spin and $\overline{1s}(P)$ to denote the electron in the $1s$ orbital with β spin. In our case this becomes

$$\frac{1}{\sqrt{N!}}|1s(1)\overline{1s}(2)2s(3)\overline{2s}(4)2p_x(5)\overline{2p}_x(6)2p_y(7)\overline{2p}_y(8)2p_z(9)\overline{2p}_z(10)| \qquad (50)$$

When an electronic state contains N electrons it follows that the antisymmetrized form of the wave function contains $N!$ terms. At first sight it appears that very rapidly there are too many terms to handle with ease. However, as we shall see, this is not the case and it is possible to write a comparatively simple expression for the energy of such a system. Before doing this, it is necessary to look first at the angular momentum properties

$$\frac{1}{\sqrt{N!}}$$

$$
\begin{vmatrix}
1s(1)\alpha(1), & 1s(2)\alpha(2), & 1s(3)\alpha(3), & 1s(4)\alpha(4), & 1s(5)\alpha(5), & 1s(6)\alpha(6), & 1s(7)\alpha(7), & 1s(8)\alpha(8), & 1s(9)\alpha(9), & 1s(10)\alpha(10) \\
1s(1)\beta(1), & 1s(2)\beta(2), & 1s(3)\beta(3), & 1s(4)\beta(4), & 1s(5)\beta(5), & 1s(6)\beta(6), & 1s(7)\beta(7), & 1s(8)\beta(8), & 1s(9)\beta(9), & 1s(10)\beta(10) \\
2s(1)\alpha(1), & 2s(2)\alpha(2), & 2s(3)\alpha(3), & 2s(4)\alpha(4), & 2s(5)\alpha(5), & 2s(6)\alpha(6), & 2s(7)\alpha(7), & 2s(8)\alpha(8), & 2s(9)\alpha(9), & 2s(10)\alpha(10) \\
2s(1)\beta(1), & 2s(2)\beta(2), & 2s(3)\beta(3), & 2s(4)\beta(4), & 2s(5)\beta(5), & 2s(6)\beta(6), & 2s(7)\beta(7), & 2s(8)\beta(8), & 2s(9)\beta(9), & 2s(10)\beta(10) \\
2p(1)\alpha(1), & 2p(2)\alpha(2), & 2p(3)\alpha(3), & 2p(4)\alpha(4), & 2p(5)\alpha(5), & 2p(6)\alpha(6), & 2p(7)\alpha(7), & 2p(8)\alpha(8), & 2p(9)\alpha(9), & 2p(10)\alpha(10) \\
2p(1)\beta(1), & 2p(2)\beta(2), & 2p(3)\beta(3), & 2p(4)\beta(4), & 2p(5)\beta(5), & 2p(6)\beta(6), & 2p(7)\beta(7), & 2p(8)\beta(8), & 2p(9)\beta(9), & 2p(10)\beta(10) \\
2p(1)\alpha(1), & 2p(2)\alpha(2), & 2p(3)\alpha(3), & 2p(4)\alpha(4), & 2p(5)\alpha(5), & 2p(6)\alpha(6), & 2p(7)\alpha(7), & 2p(8)\alpha(8), & 2p(9)\alpha(9), & 2p(10)\alpha(10) \\
2p(1)\beta(1), & 2p(2)\beta(2), & 2p(3)\beta(3), & 2p(4)\beta(4), & 2p(5)\beta(5), & 2p(6)\beta(6), & 2p(7)\beta(7), & 2p(8)\beta(8), & 2p(9)\beta(9), & 2p(10)\beta(10) \\
2p(1)\alpha(1), & 2p(2)\alpha(2), & 2p(3)\alpha(3), & 2p(4)\alpha(4), & 2p(5)\alpha(5), & 2p(6)\alpha(6), & 2p(7)\alpha(7), & 2p(8)\alpha(8), & 2p(9)\alpha(9), & 2p(10)\alpha(10) \\
2p(1)\beta(1), & 2p(2)\beta(2), & 2p(3)\beta(3), & 2p(4)\beta(4), & 2p(5)\beta(5), & 2p(6)\beta(6), & 2p(7)\beta(7), & 2p(8)\beta(8), & 2p(9)\beta(9), & 2p(10)\beta(10)
\end{vmatrix}
$$

of many-electron systems and to see how a study of such properties can lead to a systematization of the possible states which can arise from a given electronic configuration.

2.7 ANGULAR MOMENTUM OPERATORS AND THE MANY-ELECTRON ATOM

The existence of spin angular momentum defined by the operators S_x, S_y, S_z and S^2 analogous to M_x, M_y, M_z and M^2 has already been discussed. It has been mentioned that S^2 possesses only one eigenvalue $\frac{3}{4}\hbar^2$ for a single electron and the two values $\pm\frac{1}{2}\hbar$ for the eigenvalue of the component operator S_z.

The commutation rules for angular momenta were given in equation (8), Chapter 1; these are

$$M_xM_y - M_yM_x = i\hbar M_z$$
$$M_yM_z - M_zM_y = i\hbar M_x \tag{51}$$
$$M_zM_x - M_xM_z = i\hbar M_y$$

together with the commutation rules between M^2 and its components M_x, M_y and M_z.

$$M^2M_z - M_xM^2 = 0 \tag{52}$$

Because of the Uncertainty Principle only the magnitude of M^2 and one of its components M_z can be known with certainty. Knowledge of the values of M_x and M_y would involve absolute certainty about the motion of the electron. Certain relationships exist between the components M_x and M_y. These are

$$(M_x - iM_y)Y_{l,m} = \sqrt{l(l+1) - m(m-1)}\,Y_{l,(m-1)}$$
$$(M_x + iM_y)Y_{l,m} = \sqrt{l(l+1) - m(m+1)}\,Y_{l,(m+1)} \tag{53}$$

where Y is a spherical harmonic, i.e. an eigenfunction of angular momentum. The operators $(M_x - iM_y)$ and $(M_x + iM_y)$ thus decrease or increase the m quantum number by one unit respectively. When m has its lowest possible value, $(M_x - iM_y)$ annihilates the rotating particle and similarly when m has its highest value $(M_x + iM_y)$ annihilates it. These operators are often known either as step-down and step-up operators or in specific contexts as annihilation and creation operators. The two operators are often given the symbols M_- and M_+. It is interesting to relate M_+ and M_- to M^2.

$$M_+M_- = (M_x+iM_y)(M_x-iM_y) = M_x^2+M_y^2+iM_yM_x-iM_xM_y$$
$$= M_x^2+M_y^2+i(M_yM_x-M_xM_y) = M_x^2+M_y^2+i(-i\hbar M_z)$$
$$= M_x^2+M_y^2+\hbar M_z \tag{54}$$

Now $M^2 = M_x^2+M_y^2+M_z^2$, therefore substituting for $M_x^2+M_y^2$ from (54) we obtain

$$M^2 = M_+M_- -\hbar M_z+M_z^2 \tag{55}$$

Similarly it can be shown that

$$M^2 = M_-M_+ +\hbar M_z+M_z^2 \tag{56}$$

When there are several electrons in the atom considered it is possible to define a total orbital angular momentum. It will be remembered that angular momentum is a vector (defined as the vector product of linear momentum and position vectors, i.e. $r \times p$). To find the total angular momentum of the n-electron system we add together the like component for each electron and then sum the squares of the total components to find the total value of M^2. Let us define a set of angular momentum operators for each electron in the following way: M_{x1}, M_{y1} and M_{z1} are the operators for M_x, M_y and M_z for electron 1, they have eigenvalues m_{x1}, m_{y1} and m_{z1} respectively. A similar set is defined for each of the n electrons. Total M_x, M_y and M_z operators are defined in the following way

$$M_x = \sum_i M_{xi}$$
$$M_y = \sum_i M_{yi} \tag{57}$$
$$M_z = \sum_i M_{zi}$$

with

$$M^2 = M_x^2+M_y^2+M_z^2 \tag{58}$$

Since each term on the r.h.s. of (57) acts only on one electron the total x (y or z) component of angular momentum m_x is given by

$$m_x = \sum_i m_{xi}$$
$$m_y = \sum_i m_{yi} \tag{59}$$
$$m_z = \sum_i m_{zi}$$

Similar definitions give the n-electron operators $(M_x\pm iM_y)$.

When we come to consider the spin angular momentum we have a similar set of operators. Analogous to the operators M_+ and M_- we have the operators S_+ and S_- which possess the following properties

$$(S_x - iS_y)\alpha = \hbar\beta \qquad (S_x + iS_y)\alpha = 0$$
$$(S_x - iS_y)\beta = 0 \qquad (S_x + iS_y)\beta = \hbar\alpha \tag{60}$$

There is also a set of commutation relations analogous to (51)

$$S_x S_y - S_y S_x = i\hbar S_z \tag{61}$$

with similar equations between the other pairs of operators. S^2 also commutes with S_x, S_y and S_z. Analogous to (55) and (56) we have

$$S^2 = S_+ S_- - \hbar S_z + S_z^2 \tag{62}$$
$$S^2 = S_- S_+ + \hbar S_z + S_z^2 \tag{63}$$

In the n-electron case, n-electron spin operators can be defined analogous to (57), (58) and (59)

$$S_x = \sum_i S_{xi}, \qquad S_y = \sum_i S_{yi}, \qquad S_z = \sum_i S_{zi} \tag{64}$$

$$S^2 = S_x^2 + S_y^2 + S_z^2 \tag{65}$$

$$s_x = \sum_i s_{xi}, \qquad s_y = \sum_i s_{yi}, \qquad s_z = \sum_i s_{zi} \tag{66}$$

The total angular momentum of a single electron in an atomic orbital is given by the eigenvalue j of an operator J, defined by

$$J = M + S \tag{67}$$

interpreting (67) in a vector sense, viz.

$$J_x = M_x + S_x$$
$$J_y = M_y + S_y \tag{68}$$
$$J_z = M_x + M_y$$

with

$$J^2 = J_+ J_- - \hbar J_z + J_z^2$$
$$= J_x^2 + J_y^2 + J_z^2 \tag{69}$$

Similarly the eigenvalues of the components are given by

$$j_z = m_z + s_z \tag{70}$$

A set of n-electron J operators and eigenvalues can be defined in a similar way and the formulation of these is left as an exercise for the reader.

From the commutation relations (51), (52), (61) etc. and the similar set relating J^2 and its components, we can notice certain things. M^2, S^2 and J^2 commute with their components, M^2 and S^2 commute with J^2 and its components, but the components of M^2 and S^2 do not commute with either J^2 or with its components (except for like components).

It is possible to write certain sets of angular momentum operators which commute among themselves. Two such sets are M^2, M_z, S^2, S_z and J_z and the other set M^2, S^2, J^2 and J_z. Operators which commute, as seen in Chapter 1, correspond to independent observables. From a given set of angular momentum eigenfunctions (i.e. spherical harmonics $Y_{l,m}$) it is possible to construct sets which are simultaneously eigenfunctions of all of the members of one of the sets of operators given above. One important point follows from this. If, and only if, two eigenfunctions have the same eigenvalue for each of the set of commuting operators, the matrix element of the Hamiltonian between these functions is non-vanishing. If they have differing values for only one operator then the element of the Hamiltonian is zero. It is therefore worth the trouble, when faced with the problem of calculating the energy of a many-electron system, to construct linear combinations of the basis orbitals which are simultaneously eigenfunctions of one of the sets of commuting operators. By noting those with differing eigenvalues for one or more operators, we can say without further work that the matrix element of the Hamiltonian between these functions is zero.

2.8 THE VECTOR MODEL OF THE ATOM

From what was said in the last section, the behaviour of the electrons in a many-electron atom can be understood if one considers the individual electrons to be unit (or basis) vectors. The properties of angular momentum are then related to those of the individual electrons by combining the latter in vector fashion. Because of this, such an approach to the many-electron atom is known as the Vector Model of the Atom.

We now want to use this approach to predict the various states of an atom which result from a given electronic configuration. The electrons in the atom are characterized by a set of quantum numbers n_i, l_i or m_i, m_{zi} and s_{zi} for each electron. We choose to use the quantum number m for the orbital angular momentum in order to make it consistent with the terminology of the last section, and the z component of this is therefore given the symbol m_z and not m with a suffix as was used previously.

The resultant value of the orbital angular momentum is given by

$$m = \sum_i m_i \tag{71}$$

The resultant value will take values from $+m_i$ through units of one down to the lowest possible positive number. For example, if we consider an atom with one d electron and one p electron, then the possible values for m are $2+1 = 3, 2+0 = 2$ and $2-1 = 1$ depending upon whether the vectors are aligned parallel, perpendicular, or antiparallel to each other.

The resultant values of m_z are obtained by adding together the individual m_z's in an arithmetic way, since they are all components of the same unit vector. For instance, in the example given above, the d electron can have m_z values of $2, 1, 0, -1, -2$ and the p electron values $1, 0, -1$, the resultant will contain values from $+3$ to -3 with some of the intermediate values occurring more than once.

The resultant values of s_z are obtained from the possibilities in the same way as the values of m_z.

We can only understand how this works out by taking an example. One important point to note before we do this, is that the contribution to m_z and s_z from filled shells and subshells is zero. This can be easily seen by looking at any example. In constructing the vector model for any system we need only consider the electrons in the valence shell, i.e. the unfilled shell or subshell.

Consider the case of an atom which has two electrons in the valence shell, one of p type and one of d type.

The orbital angular momentum for the two electrons is

$$m = m_p + m_d$$

$m_p = 1$ and $m_d = 2$, the possible m values are 3, 2 and 1. Many electron states with $m = 0$ are called "S states" (by analogy with the s orbital for the single electron), those with $m = 1$ "P states" and with $m = 2$ "D states" and so on. The possible states arising from the configuration pd are F ($m = 3$), D and P. We cannot as yet say with what multiplicity these states occur or indeed how many of them there are.

Next we look at the possible values of m_z. For the p electron we have $m_{zp} = +1, 0$ and -1 as possible values and for the d electron $m_{zd} = +2, +1, 0, -1$ and -2. To obtain the possible values of m_z from these it is most convenient to set them out in the form of a table as shown on page 53. From the table of resultant m_z values it is seen that $+3$ and -3 each occur once, $+2$ and -2 each occur twice, $+1$ and -1 each occur four times and 0 occurs four times.

$$m_{zd}$$

+2	+1	0	−1	−2		
+3	+2	+1	0	−1	+1	
+2	+1	0	−1	−2	0	m_{zp}
+1	0	−1	−2	−3	−1	

In order to allocate these into F, D, P and S terms it is necessary to construct the possible s_z values. Each electron can have s_z values of $+\frac{1}{2}$ and $-\frac{1}{2}$ and the possible resultants are $+1$, 0 (twice) and -1. The value $s_z = 0$ occurs twice because the three other one-electron quantum numbers are not the same and therefore $s_{zp} = +\frac{1}{2}$, $s_{zd} = -\frac{1}{2}$ and $s_{zp} = -\frac{1}{2}$, $s_{zd} = +\frac{1}{2}$ represent two distinguishable spin states. All four values of the spin can be associated with each value of m_z.

In writing down the possible combinations of the m_z values it is necessary to notice one or two points. The necessary values of s_z to form a triplet are 1, 0 and -1, to form a doublet $+\frac{1}{2}$ and $-\frac{1}{2}$. This arises because the members of a multiplet have s_z values ranging from $s, s-1, s-2, \ldots, 0$ to $-1, -2, \ldots, -s$, in other words $2s+1$ members for a given s value. In other words the same relation exists between s and s_z as exists between m and m_z. A rule known as "Hund's rule" states that where a choice is possible the state of maximum multiplicity has the lowest energy, i.e. for a given m a triplet lies lower in energy than a singlet, a quartet lower than a doublet, a quintet lower than a triplet and so on.

On the bases of these we construct the possible atomic states for our two-electron system.

Starting with the highest possible m value, i.e. 3, we take from the table of m_z values

$$+3 \quad +2 \quad +1 \quad 0 \quad -1 \quad -2 \quad -3$$

which leaves $m_z = \pm 2$ once, $m_z = \pm 1$ and 0 three times each. For each of the m_z values in the $m = 3$ term, each of the four possible spin values occurs. This gives rise to a triplet ($s_z = 1, 0-1$) and a singlet (from the remaining $s_z = 0$). The states ^3F and ^1F exist for the configuration pd.

From the remaining terms of m_z, states with $m = 2$ (D states) can be found. For this the values

$$+2 \quad +1 \quad 0 \quad -1 \quad -2$$

are required. This leaves $m_z = \pm 1, 0$ each occurring twice. The D state also occurs both as a triplet and a singlet. From the remaining m_z terms, two P states can be formed. One of these occurs as a triplet and the other as a singlet.

The possible states which arise from an electronic configuration are

$$^3F, \, ^1F, \, ^3D, \, ^1D, \, ^3P, \, ^1P$$

Some other features of interest arise when the electrons have the same n, m values, e.g. the configuration $(np)^2$. Here it is possible to have D, P and S terms. The table of possible m_z is

$$m_{z1}$$

+1	0	−1	
+2	+1	0	+1
+1	0	−1	0
0	−1	−2	−1

m_{z2}

Since $n_1 = n_2$ and $m_1 = m_2$, the m_z values $+1$ which arise from $m_{z1} = 1, m_{z2} = 0$ and from $m_{z1} = 0, m_{z2} = 1$ do not belong to distinguishable states and therefore only one occurs. Similar arguments apply to the off-diagonal 0 and -1 values in the table. The values on the lower side of the diagonal can be ignored whenever two equivalent electrons are considered. We are therefore left with

$$\begin{array}{ccc} +2 & +1 & 0 \\ & 0 & -1 \\ & & -2 \end{array}$$

When considering the possible s_z values which can be combined with each m_z, it is necessary to notice that for each term occurring along the diagonal, $n_1 = n_2, m_1 = m_2, m_{z1} = m_{z2}$. Therefore by the Pauli principle $s_{z1} \neq s_{z2}$ and in this case $s_{z1} = \frac{1}{2}, s_{z2} = -\frac{1}{2}$ and $s_{z1} = -\frac{1}{2}, s_{z2} = \frac{1}{2}$ do not correspond to distinguishable states and so $s_z = 0$ only occurs once. In the off-diagonal cases $s_z = 1, 0$ (twice) and -1 can all be combined with each m_z value. The possible combinations are

$$\begin{array}{ccc} +2(0) & +1(1, 0, 0, -1) & 0(1, 0, 0, -1) \\ & 0(0) & -1(1, 0, 0, -1) \\ & & +2(0) \end{array}$$

where the possible s_z values have been placed in brackets behind the m_z value.

From the components given in the table it is possible to form the following states ^1D, ^3P, ^1P and ^1S. When we consider the states arising from equivalent electrons, Hund's rules tell us that where we have states with differing m values but the same multiplicity, the state with highest m is lowest. When the shell or subshell is more than half-full Hund's rules are inverted and the state with lowest multiplicity is the lowest in energy, also the state with lowest m is the lowest among states with the same multiplicity.

It has been seen that the possible spectroscopic states which arise from a given electronic configuration can be written down without any lengthy calculation, but this is not so for the evaluation of the energy of these states, and our next task is to derive the energy expression and hence the Hamiltonian for a many-electron atom.

2.9 THE ENERGY OF THE MANY-ELECTRON ATOM

Consider an atom with a closed-shell configuration. The ground state of such a system can be well represented by a single determinant. Let this determinant be

$$\Psi = \frac{1}{\sqrt{N!}} |\phi_a(1)\alpha(1)\Phi_a(2)\beta(2)\phi_b(3)\alpha(3)\cdots\phi_n(N)\beta(N)| \qquad (72)$$

The Hamiltonian for the N-electron atom is

$$h = \sum_i \left(-\frac{1}{2m}\nabla_i^2 - \frac{Ze^2}{r_i} \right) + \frac{1}{2}\sum_{i,j}\frac{e^2}{r_{ij}} \qquad (73)$$

where the term $\sum_i -Ze^2/r_i$ describes the interaction of all the electrons with the nucleus, and the term $\sum_{i,j} 1/r_{ij}$ represents the interaction of all the electrons with each other, and this is given the factor $\frac{1}{2}$ in (73) to ensure that the interaction between a given pair of electrons is counted only once. The first term in (73) refers to the kinetic energy of the N electrons.

Using (72) and (73) the wave equation

$$h\Psi = E\Psi \qquad (74)$$

is obtained.

If the interaction between the electrons is ignored, (74) becomes separable into N equations each of which is a function of only one electron. It is this fact that justifies the use of the one-electron orbital product wave function. The form of the orbitals in (72) often includes a variational parameter to allow the stationary value of the energy to be found. The energy which corresponds to (74) is found in the usual way.

$$E = \int \Psi^* h \Psi \, d\tau = \int \Psi^* \left(\sum_i -\frac{1}{2m} \nabla_i^2 \right) \Psi d\tau$$

$$- \int \Psi^* \left(\sum_i \frac{Ze^2}{r_i} \Psi \, d\tau + \frac{1}{2} \int \Psi^* \sum_{ij} \frac{e^2}{r_{ij}} \Psi \right) d\tau \qquad (75)$$

The evaluation of each integral on the r.h.s. must be considered separately. Consider first the kinetic energy integral:

$$T = \frac{1}{N!} \int |\phi_a^*(1)\alpha^*(1) \cdots \phi_n^*(N)\beta^*(N)| \left(\sum_i -\frac{1}{2m} \nabla_i^2 \right) |\phi_a(1)\alpha(1) \cdots \phi_n(N)\beta(N)| \, d\tau$$

$$(76)$$

As it stands there are $N!^2$ terms in this integral as the determinantal wave function contains $N!$ terms. However it is not necessary to evaluate this large number of one-electron integrals. Examine the integral with respect to electron (1) in (76) and in particular the one with the leading term on the l.h.s. of the integrand. This integral is

$$\frac{1}{N!} \int \phi_a^*(1)\alpha^*(1) \cdots \phi_n^*(N)\beta^*(N) \left(\frac{-1}{2m} \nabla_1^2 \right)$$

$$\times |\phi_a(1)\alpha(1) \cdots \phi_n(N)\beta(N)| \, d\tau_1 \, d\tau_2 \, d\tau_3 \cdots d\tau_n \qquad (77)$$

The integration over the other $N-1$ electrons can be factored out and (77) becomes

$$\frac{1}{N!} \int \phi_a^*(1)\alpha^*(1) \left(-\frac{1}{2m\nabla_1^2} \right) \phi_a(1)\alpha(1) \, d\tau_1 \int \phi_a^*(2)\beta(2) \cdots \phi_n^*(N)\beta^*(N)(-1)^p$$

$$\times P[\phi_a(2)\beta(2)\phi_b(3)\alpha(3) \cdots \phi_n(N)\beta(N)] \, d\tau_2 \, d\tau_3 \cdots d\tau_n \qquad (78)$$

Of all the terms in the second integral the only one which is non-vanishing is the one which is identical with the l.h.s. (assuming the orbitals to be orthogonal). The factor to multiply the kinetic energy integral in (78) becomes one, since it is only the term on the r.h.s. which is identical to that on the left which is non-vanishing. Similarly for the other terms in the determinantal wave function on the l.h.s. in (76). The value of the integral is characterized by the orbital and not by the electron, which here becomes a dummy suffix. There will be $N!$ occurrences of the integral $\int \phi_a^*(X)(-\frac{1}{2}\nabla_x^2)\phi_a(X) \, dv_x$ in (76). This is the case for each orbital. The value of (76) is

$$T = \frac{1}{N!} \cdot N!(T_a + T_a + T_b + T_b + \cdots + T_n) \tag{79}$$

or, on collecting like terms together and simplifying

$$T = 2 \sum_{i=1}^{n} T_i \tag{80}$$

A similar reasoning applies to the other one-centre integrals V_i which describe the attraction between electron i and the nucleus. The total electron–nucleus interaction is

$$V = 2 \sum_i V_i \tag{81}$$

where again the suffix refers to the orbital and not to the electron.

In writing down the energy the two one-centre integrals are usually combined together to give a total one-electron contribution to the energy f_i

$$f_i = T_i + V_i \tag{82}$$

and the total one-electron contribution to the energy is

$$f = 2 \sum_i f_i \tag{83}$$

We now consider the evaluation of the two-electron integrals in (75) which represent the interaction between the electrons. As with the one-electron integrals, examine the contribution from the leading term on the l.h.s.

$$\int \phi_a^*(1)\alpha^*(1)\phi_a^*(2)\beta^*(2) \cdots \phi_n^*(N)\beta^*(N) \sum_{XY} \frac{e^2}{r_{XY}} |\phi_a(1)\alpha(1) \cdots \phi_n(N)\beta(N)| \, d\tau \tag{84}$$

Examine the value of the integral representing interaction between electron 1 and electron 2. The relevant term is

$$\int \frac{\phi_a\alpha(1)\phi_a(2)\beta(2)\phi_a(1)\alpha(1)\phi_a(2)\beta(2)}{r_{12}} \, d\tau_1 \, d\tau_2$$

$$\times \int \phi_b^*(3)\alpha(3) \cdots \phi_n^*(N)\beta(N)(-1)^P P[\phi_b(3)\alpha(3) \cdots \phi_n(N)\beta(N)] \, d\tau_3 \cdots d\tau_N \tag{85}$$

As in the one-electron case, the only contribution to (85) comes from the term in the second integral which has both sides of the integrand matching. Similarly there is one such term from each member of the wave function of which the leading term is given in (84). Again the value of the term is characteristic of the orbital and not the electron and we give it the symbol J_{aa}.

Examine the contribution from the interaction between electrons 1 and 3 in (84). This represents the interaction between one electron in orbital ϕ_a and one electron in orbital ϕ_b. The integral is given the symbol J_{ab}. A similar J_{ab} arises from the interaction of electron 1 and 4, a third from the interaction of electrons 2 and 3, and a fourth from electrons 2 and 4. This makes a total contribution of $4J_{ab}$. This interaction is the quantum-mechanical analogue of the Coulombic interaction between two charge clouds.

There is however another type of two-electron interaction. Consider the integral analogous to (85) involving $1/r_{13}$. This can be represented as a product of two integrals, one over space and one over spin. This gives

$$\int \frac{\phi_a^*(1)\phi_b^*(3)\phi_a(1)\phi_b(3)}{r_{13}} \, dv_1 \, dv_3 \int \alpha^*(1)\alpha^*(3)\alpha(1)\alpha(3) \, d\eta_1 \, d\eta_3 \qquad (86)$$

Interchanging electrons 1 and 3 on the r.h.s. of each integral we obtain

$$\int \frac{\phi_a^*(1)\phi_b^*(3)\phi_a(3)\phi_b(1)}{r_{13}} \, dv_1 \, dv_3 \int \alpha^*(1)\alpha^*(3)\alpha(3)\alpha(1) \, d\eta_1 \, d\eta_3 \qquad (87)$$

The spin integral still integrates out to unity, but the space integral gives something which has no classical analogue. It is known as a resonance integral and given the symbol K_{ab}. Its contribution to the energy has a negative sign because in forming the integral we interchange two particles, which implies by application of the Pauli principle that the wave function changes sign. There is a further contribution K_{ab} which arises from electrons 2 and 4. Resonance or exchange integrals only arise when the particles concerned have parallel spins, otherwise the spin integral has the value zero. In the closed shell situation therefore there are only half as many exchange integrals as there are Coulombic integrals. The two-electron contribution to the energy is

$$J_{ii} + \tfrac{1}{2} \sum_{ij} (4J_{ij} - 2K_{ij}) \qquad (88)$$

which simplifies further if we define a quantity $K_{ii} = J_{ii}$ and simplify the bracket, which gives

$$\sum_{ij} (2J_{ij} - K_{ij}) \tag{89}$$

The total energy of the N-electron atom is obtained by adding (83) and (87) together to give

$$E = 2\sum_{i} f_i + \sum_{ij} (2J_{ij} - K_{ij}) \tag{90}$$

This expression for the energy of a many-electron atom was first derived by J. C. Slater.

It can be seen from the form of the integrals J_{ij} and K_{ij} that these integrals are dependent on the charge distribution. The energy is then a function of the charge distribution, and in calculating the wave function and energy of a many-electron atom, it is necessary to vary the wave function until energy and charge distribution are self-consistent, using the usual variation method for each cycle. This is often a long and laborious process.

The best wave function obtained from this procedure is, however, still only an approximation to the true wave function. A better wave function can be obtained by setting aside the one-determinant wave function and writing the ground state wave function as a linear sum of determinants,

$$\Psi_0 = \sum_{i=0}^{N} c_i \Phi_i \quad \text{(where we write } \Phi_i \text{ for a single determinant)} \tag{91}$$

Provided that the configurations selected have the same symmetry, they will interact with each other, and the resultant wave function with the lowest energy will be the best approximation to the ground state. Provided that a sufficient number of determinants are chosen, the energy will be lower, and thus a better approximation to the ground state energy than the one-determinant approximation. It follows that the wave function obtained will be a better approximation to the true wave function than the one-determinant wave function. One advantage of this approximation is that the number of terms in the wave function can be extended at will.

2.10 THE SPECTRUM OF AN ATOM

An atom may absorb radiation and become excited. One of the electrons absorbing this energy will occupy an orbital of higher energy than it occupies in the ground state. In the Bohr theory the transition energy, corresponding to the energy absorbed, is obtained by evaluating a simple difference in orbital energies:

$$E_{i \to j} = E_j - E_i \qquad (92)$$

where E_j and E_i are the energies of the excited state orbital ψ_j and of the ground state orbital ψ_i respectively. Following through the techniques of the last section this is found to be no more than an approximation to the true transition energy. It is true only when electron interactions are ignored.

Consider an N-electron atom whose wave function can be represented as a single determinant. In an excited state one electron has been excited from orbital j to orbital l. There are four ways in which the wave function can be represented, depending upon the way in which the spin-eigenfunction for the excited state is written. The four possible determinants are:

$$\Phi_A = |\psi_1(1)\alpha(1)\psi_1(2)\beta(2) \cdots \psi_j(L)\alpha(L)$$

$$\cdots \psi_{N/2}(N-1)\alpha(N-1)\psi_{N/2}(N)\beta\beta(N)\psi_l(L+1)\beta(L+1)|$$

$$\Phi_B = |\psi_1(1)\alpha(1)\psi_1(2)\beta(2) \cdots \psi_j(L)\beta(L)$$

$$\cdots\cdots\cdots\cdots\cdots\cdots\cdots\cdots\cdots\cdots\cdots\cdots \psi_l(L+1)\alpha(L+1)|$$

$$\Phi_C = |\psi_1(1)\alpha(1) \cdots\cdots\cdots\cdots \psi_j(L)\alpha(L)$$

$$\cdots\cdots\cdots\cdots\cdots\cdots\cdots\cdots\cdots\cdots\cdots\cdots\cdots \psi_l(L+1)\alpha(L+1)| \qquad (93)$$

$$\Phi_D = |(\psi_1(1)\alpha(1)\cdots\cdots\cdots\cdots \psi_j(L)\beta(L)$$

$$\cdots\cdots\cdots\cdots\cdots\cdots\cdots\cdots\cdots\cdots\cdots\cdots \psi_l(L+1)\beta(L+1)|$$

On examining these four wave functions it is found that the first two of these eigenvalues of S_z equal to zero and the other two have values of plus and minus one respectively. We take linear combinations of Φ_A and Φ_B, viz:

$$\Phi_{A'} = \frac{1}{\sqrt{2}}(\Phi_A - \Phi_B)$$

$$\Phi_{B'} = \frac{1}{\sqrt{2}}(\Phi_A + \Phi_B) \qquad (94)$$

It is found that $\Phi_{A'}$ is an eigenfunction of S^2 with eigenvalue zero and $\Phi_{B'}$ and Φ_C and Φ_D of (93) are eigenfunctions of S^2 with eigenvalues

$2h^2/4\pi^2$. The wave functions correspond to a singlet and triplet state respectively. The wave functions for the singlet and triplet are

Singlet $$^1\Psi_{j\to l} = \frac{1}{\sqrt{2}}(\Phi_A - \Phi_B) \tag{95}$$

Triplet $$^3\Psi_{j\to l} = \begin{cases} \dfrac{1}{\sqrt{2}}(\Phi_A + \Phi_B) \\ \Phi_C \\ \Phi_D \end{cases} \tag{96}$$

The energy expression for these states can be evaluated following the methods of the last section, and it is found that these are

$$^{1,3}E_{j\to 1} = 2\sum_i f_i + f_j + f_l + \sum_{h,i}(2J_{hi} - K_{hi}) + \sum_i(2J_{ij} - K_{ij})$$
$$+ \sum_i(2J_{il} - K_{il}) + J_{jl} \pm K_{jl} \tag{97}$$

where we take the $+$ sign for the singlet and the $-$ sign for the triplet. The orbital energies E_j and E_l (ground and excited state orbitals respectively) are given by

$$E_j = f_j + \sum_i(2J_{ij} - K_{ij})$$
$$E_l = f_l + \sum_i(2J_{il} - K_{il}) \tag{98}$$

where the summation is over orbitals occupied in the ground state only. With these definitions (97) becomes

$$^{1,3}E_{j\to l} = E_l - E_j - J_{jl} + K_{jl} \pm K_{jl} \tag{99}$$

where we take $+$ for the singlet and $-$ for the triplet. From (99) it can be seen that the transition energy is not a simple difference in orbital energies but contains other two-electron terms.

In addition it must be realized that the excited states whose energies are given by (99) are only those involved in transitions between the two orbitals j and l. A better excited state wave function is found when several configurations with the same multiplicity and symmetry, but involving different transitions, are allowed to interact. These many-configuration wave functions give better approximations to the excited state under consideration. The dominant term in this wave function will, as a general rule, be the configuration describing the orbital jump under consideration.

3

Diatomic Molecules

3.1 INTRODUCTION

The essential difference between atoms and molecules, as far as their electronic structures are concerned, is that the monocentric atom is now replaced by an electronic system in which there is more than one nucleus. This raises many new problems. In the case of the atom, the electronic motion was referred to the nucleus as origin, and since this is the centre of mass, the process is legitimate. When we go to the molecule, however, this is no longer the case. The centre of mass is no longer the nucleus since there are now several nuclei. The Hamiltonian for the system is

$$H = -\frac{1}{2}\sum_{\alpha}\frac{1}{m_\alpha}\nabla_\alpha^2 - \frac{1}{2}\sum_{j}\frac{1}{m_j}\nabla_j^2 - \sum_{\alpha,j}\frac{Z_\alpha}{r_{\alpha j}} + \sum_{\alpha,\beta}\frac{Z_\alpha Z_\beta}{R_{\alpha\beta}} + \sum_{ij}\frac{1}{r_{ij}} \qquad (1)$$

$$\underbrace{\phantom{-\frac{1}{2}\sum_{\alpha}\frac{1}{m_\alpha}\nabla_\alpha^2}}_{\text{(all nuclei)}} \underbrace{\phantom{-\frac{1}{2}\sum_{j}\frac{1}{m_j}\nabla_j^2}}_{\text{(all electrons)}}$$

Provided that there are no interactions between the electrons and the nuclei, a wave function which is a simple product of a nuclear and an electronic wave function can be assumed, viz.

$$\Psi = \psi_{\text{nuc.}}\psi_{\text{elect.}} \qquad (2)$$

This leads to a separation of variables, and the two wave equations which result can be solved independently. This procedure can be justified on physical grounds. The nuclei are so many times heavier than the electrons that the electronic motions will occur much more rapidly than the nuclear motions, and furthermore because of their greater mass the nuclei will be unable to "follow" the electronic motion. The nuclei are therefore regarded as fixed. This assumption is known as the Born–Oppenheimer approximation. In practice this allows us to drop the nuclear kinetic energy terms from the Hamiltonian.

The complete Hamiltonian requires us to include the various nucleus–electron interactions, but in practice these terms are assumed to be negligible. A certain amount of work has been carried out recently to test the validity of the Born–Oppenheimer approximation.

Having introduced this considerable simplification the Hamiltonian is still insoluble. Except in the simplest case we are still left with a three-body problem which cannot be solved in closed form. In order to simplify the problem, some further assumption about the form of the wave function must be made. It is easy to find approximate molecular wave functions which can be used as variation functions, and a "best" molecular wave function obtained using variation theory. It may seem at first sight that molecular calculations could be approached through some form of perturbation theory. However, this is not the case since the perturbation involved in every case is far too large to be amenable to any such approach.

Attempts to solve molecular problems fall into two distinct categories. First, the molecular orbital theory, in which the electrons are assigned to orbitals which are analogous to the atomic orbitals in the atomic case, only these orbitals are now polycentric instead of monocentric as in the atomic case. The other approach is along the lines of the conventional valency bonds of chemical valence theory. Here chemical intuition is used in assigning the electrons to certain "electron pair" bonds. Many-electron valence pair functions are allowed to interact using a variational approach, leading to a wave function which can be represented as a linear sum of functions in which perfect pairing occurs. Both theories have their merits and under perfect conditions would lead to the same final wave function, if this was the "true" wave function.

3.2 THE LINEAR COMBINATION OF ATOMIC ORBITALS

It is important to find the right kind of variation function for the molecular orbital theory. Consider an array of atoms

$$\cdot \quad \cdot \quad \cdot \quad \cdot \quad \cdot \quad \cdot \quad \cdot$$
$$a \quad b \quad c \quad d \quad e \quad f \quad g$$

Assume that each atom has available for bonding one electron derived in each case from the same type of atomic orbital, e.g. each atom has available one $2p$ electron. If the interatomic distance is large each atom will behave as if it is independent of the others. As we bring the nuclei together the charge clouds on adjacent atoms will overlap leading to a many-centred delocalized orbital. Such a state can be represented by a linear function of the type

$$\psi_i = \sum_j c_{ij}\phi_j \tag{3}$$

in which the ϕ's are the atomic orbitals and the c_i are coefficients whose values are to be determined and so that the complete set is normalized. Such a function is identical with the linear variation function [equation (116), Chapter 1] whose properties have already been discussed.

If the interatomic distance is great then all of the diagonal elements of the Hamiltonian H_{ii} are given by the Hamiltonian for the isolated atom H_0. All overlap integrals S_{ij} are zero and there is no interaction in the system. As we bring the atoms closer together the electron on one atom commences to interact both with the adjacent nucleus/nuclei and with the electrons in neighbouring orbitals. The diagonal elements of H are no longer given by the element for the isolated atom but are modified by the neighbouring interactions. The clouds overlap, leading to non-vanishing overlap integrals S_{ij} as well as non-vanishing elements H_{ij}.

Following the method of section 2.5, the energy has a stationary value when each and every $\partial E/\partial c_i = 0$, and this requires that

$$\sum_j (H_{ij} - S_{ij})c_j = 0 \qquad \text{(for } i = 1, 2, \ldots, n) \tag{4}$$

This set of equations has a non-trivial solution only when

$$\begin{vmatrix} (H_{11} - E) & (H_{12} - ES_{12}) \cdots (H_{1n} - ES_{1n}) \\ (H_{21} - ES_{21}) & (H_{22} - E) \\ \vdots \\ (H_{n1} - ES_{n1}) & \qquad\qquad\qquad (H_{nn} - E) \end{vmatrix} = 0 \tag{5}$$

The eigenvalues E_k of this determinant are then the energies of the available molecular orbitals. The set of coefficients c_j for each value E_k is determined by substitution of E_k into (4) and normalizing the resulting set of coefficients. When the C_{jk} have been determined in this way, the linear variation function becomes a molecular orbital with energy E_k. Systematic substitutions of all of the eigenvalues leads to the complete set of molecular orbitals for the case considered.

Having determined the orbitals and their orbital energies, the available electrons are then fed into the orbitals two at a time starting with the orbital of lowest energy. This procedure leaves a number of vacant orbitals. It will be seen later that these orbitals can be used to gain a knowledge of the excitation energies of the molecule, and the orbital coefficients yield important information on the electron distribution in the excited state.

This method of forming molecular orbitals is known as the linear combination of atomic orbitals (LCAO) approximation.

3.3 THE HYDROGEN MOLECULE ION

The simplest molecular system is that which contains only one electron moving in the field of two hydrogen nuclei. An approximation to the lowest energy molecular orbital of this molecule can be obtained by considering a linear variation function built out of the $1s$ orbitals on each hydrogen atom.

Designating the nuclei by the letters a and b, the $1s$ orbital of an electron moving in the field of a is given the symbol ϕ_a, and that for the electron moving in the field of b the symbol ϕ_b. From the hydrogen wave functions given in the last chapter, we have

$$\phi_a = \frac{1}{\sqrt{\pi}} \exp(-r_a)$$

$$\phi_b = \frac{1}{\sqrt{\pi}} \exp(-r_b)$$

(6)

where the distances r_a and r_b are expressed in atomic units (the radius of the Bohr $1s$ orbit for hydrogen).

The LCAO molecular orbitals will have the form

$$\psi_J = c_{aJ}\phi_a + c_{bJ}\phi_b \tag{7}$$

The coefficients c_a and c_b must be determined from a variation calculation. The relevant matrix elements are defined by

$$H_{aa} = \int \phi_a H \phi_a \, d\tau = H_{bb} = \int \phi_b H \phi_b \, d\tau$$

$$H_{ab} = H_{ba} = \int \phi_a H \phi_b \, d\tau \qquad S = \int \phi_a \phi_b \, d\tau$$

The two simultaneous equations from which the stationary values of the energy are determined are

$$(H_{aa} - E)c_a + (H_{ab} - ES)c_b = 0$$

$$(H_{ba} - SE)c_a + (H_{bb} - S)c_b = 0$$

(8)

The only non-trivial solution arises when

$$\begin{vmatrix} H_{aa} - E & H_{ab} - ES \\ H_{ba} - SE & H_{bb} - ES \end{vmatrix} = 0 \tag{9}$$

Since $H_{aa} = H_{bb}$, the roots of this determinant are given by

$$H_{aa} - E = \mp(H_{ab} - SE) \tag{10}$$

or

$$E_1 = (H_{aa} + H_{ab})/(1 + S) \quad \text{and} \quad E_2 = (H_{aa} - H_{ab})/(1 - S) \tag{11}$$

E_1 will be lower in energy than E_2 if the molecular species is relatively stable. Substitution of E_1 into (9) leads to

$$(H_{aa}S - H_{ab})c_a + (H_{ab} - SH_{aa})c_b = 0 \tag{12}$$

from which it can be seen that $c_a = c_b$. In order that the variation function be normalized we note that

$$\int \psi_1 \psi_1 \, d\tau = c_a^2 + c_b^2 + 2c_a c_b S = 1$$

from which

$$c_a = c_b = \frac{1}{(2 + 2S)^{\frac{1}{2}}}$$

Substitution of the second root E_2 into (9) gives

$$c_a = -c_b = \frac{1}{(2 - 2S)^{\frac{1}{2}}}$$

The wave functions and their energies are given by

$$\psi_1 = \frac{1}{(2 + 2S)^{\frac{1}{2}}}(\phi_a + \phi_b) \qquad E_1 = \frac{H_{aa} + H_{ab}}{1 + S}$$

$$\psi_2 = \frac{1}{(2 - 2S)^{\frac{1}{2}}}(\phi_a - \phi_b) \qquad E_2 = \frac{H_{aa} - H_{ab}}{1 - S}$$

In this particular case it is possible to write down the molecular orbitals and also the symbolic form of the orbital energies without an explicit knowledge of the Hamiltonian. Usually this is impossible. It is possible in this case only because the two orbitals are determined by the symmetry of the problem.

In order to make any further progress it is necessary to formulate the Hamiltonian properly and to evaluate the matrix elements.

The Hamiltonian for the hydrogen molecule ion is

$$H = -\left(\tfrac{1}{2}\nabla^2 + \frac{1}{r_a} + \frac{1}{r_b} - \frac{1}{R}\right) \tag{13}$$

where the first term is the kinetic energy of the electron, the second term represents the interaction of the electron with nucleus a, the third term the interaction with nucleus b and the last term the repulsion between the two nuclei (all distances expressed in atomic units). Since we are using hydrogen $1s$ wave functions we have

$$H\phi_a = \left(-\tfrac{1}{2}\nabla^2 - \frac{1}{r_a}\right)\phi_a - \left(\frac{1}{r_b} - \frac{1}{R}\right)\phi_a$$

$$= \left(E_H - \frac{1}{r_b} + \frac{1}{R}\right)\phi a$$

where E_H is the energy of the hydrogen $1s$ orbital.

The matrix elements of the Hamiltonian then become

$$H_{aa} = E_H + \frac{1}{R} - \varepsilon_{aa} \qquad \text{where} \quad \varepsilon_{aa} = \int \frac{\phi_a^2}{r_b}\,d\tau$$

$$H_{ab} = \left(E_H + \frac{1}{R}\right)S - \varepsilon_{ab} \quad \text{where} \quad \varepsilon_{ab} = \int \frac{\phi_a\phi_b}{r_a}\,d\tau$$

Substituting these values into (11), we have

$$E_1 = E_H + \frac{1}{R} - \frac{\varepsilon_{aa} + \varepsilon_{ab}}{1+S} \quad \text{and} \quad E_2 = E_H + \frac{1}{R} - \frac{\varepsilon_{aa} - \varepsilon_{ab}}{1-S} \tag{14}$$

It remains now to evaluate the overlap integral S and the two interaction integrals ε_{aa} and ε_{ab}. These are evaluated by transforming the variables to elliptical coordinates. The transformation is as follows:

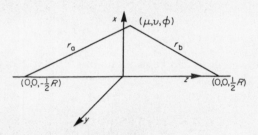

FIGURE 4.

$$x = \tfrac{1}{2}R\{(\mu^2 - 1)(1 - v^2)\}^{\frac{1}{2}}\cos\phi$$

$$y = \tfrac{1}{2}R\{(\mu^2 - 1)(1 - v^2)\}^{\frac{1}{2}}\sin\phi$$

$$z = \tfrac{1}{2}R\mu v$$

$$\mu = \frac{r_a + r_b}{R}$$

and

$$v = \frac{r_a - r_b}{R}$$

and the volume element

$$d\tau = \frac{R^3}{8}(\mu^2 - v^2)\,d\mu\,dv\,d\phi$$

where the variables have the ranges $1 \leqslant \mu \leqslant \infty$; $-1 \leqslant v \leqslant 1$; $0 \leqslant \phi \leqslant 2\pi$.
The overlap integral S is given by

$$S = \int \phi_a\phi_b\,d\tau = \frac{1}{\pi}\int \exp(-r_a - r_b)\,d\tau$$

$$= \frac{R^3}{8}\int_1^\infty \exp(-R\mu)\,d\mu \int_{-1}^{+1}(\mu^2 - v^2)\,dv \int_0^{2\pi}d\phi$$

$$= \frac{R^3}{2}\int_1^\infty \mu^2\exp(-R\mu)\,d\mu - \frac{R^3}{6}\int_1^\infty \exp(-R\mu)\,d\mu$$

These integrals are both included in the form

$$\int_1^\infty x^n\exp(-ax)\,dx = \frac{n!\exp(-a)}{a^{n+1}}\sum_{k=0}^n \frac{a^k}{k!}$$

The overlap integral then becomes

$$S = \exp(-R)\left(1 + R + \frac{R^2}{3}\right) \tag{15}$$

The two other integrals can be evaluated by similar procedures giving

$$\varepsilon_{aa} = \frac{1}{R}\{1 - (1 + R)\exp(-2R)\} \quad \text{and} \quad \varepsilon_{ab} = (1 + R)\exp(-R) \tag{16}$$

When R is large it can be seen that S approaches zero, H_{aa} approaches E_H and H_{ab} approaches zero; thus both E_1 and E_2 tend to E_H, the energy of the $1s$ orbital of hydrogen. This is exactly as one would expect. When we examine the behaviour as R tends to zero however, we find that in the limit we do not obtain the correct value for He^+. This is hardly surprising since we would never expect the LCAO approximation to hold at such small distances.

It is instructive to examine the variation of electron density with internuclear distance. For the orbital ψ_1, the electron density P_1 is given by

$$P_1 = \frac{1}{2+2S}(\phi_a^2 + \phi_b^2 + 2\phi_a\phi_b) \tag{17}$$

Mid-way between the nuclei, P_1 has the value $4/(2+2S)$. It rises to a maximum in the vicinity of the nuclei and then decreases to zero on the sides remote from the internuclear region.

For the orbital ψ_2, P_2 drops to zero mid-way between the nuclei.

The complete plot for P is given below

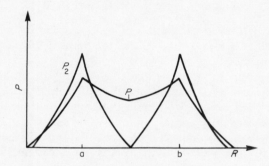

FIGURE 5. Electron density for H_2^+ orbitals.

An examination of (14) shows that the electron in state ψ_1 is energetically more stable than if it was isolated on one hydrogen atom in the field of the other; similarly an electron in state ψ_2 is in an energetically unfavourable state. The electron density plots show that P_1 has a substantial electron density between the nuclei whilst P_2 is electronically deficient in the internuclear region. The lower energy state leads to a stable molecular entity and is bonding whilst the higher one is an antibonding state. It is valuable to plot the energy of the electron in both states compared with the energy of an isolated hydrogen atom. Here, it can be easily seen that state ψ_1 leads to a stable molecule whilst the higher state is energetically unstable with respect to the isolated hydrogen atom. (See figure 6.)

The depth of the minimum in the energy curve is known as the dissociation energy D_e and corresponds to the amount of energy which must be added to the system to break the bond. A large value of D_e leads to a very stable molecule whilst a low value of D_e indicates that the molecule is not very stable. In the case of the hydrogen molecule ion, D_e is found to

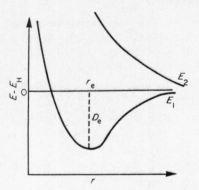

FIGURE 6. Potential energy curve for H_2^+.

have the value 1·76 eV at the equilibrium internuclear distance of 1·32 Å. Experimentally it is found that D_e has the value 2·791 eV at the internuclear distance 1·06 Å.

It is hardly surprising that the agreement is not quantitative since we have taken the smallest possible number of orbitals out of which to construct our LCAO molecular orbitals. Better agreement with experiment is obtained by using more orbitals, but the best results are obtained by discarding hydrogen-like orbitals completely and using a variation function containing many parameters. When this is done it is possible to get to within about 0·02 eV of the experimental value at the correct internuclear distance. Such calculations were carried out by James and Coolidge in the 1930's and have since been improved by several authors.

Consideration of the hydrogen molecule ion shows clearly the method of molecular orbital approach to the calculation of the electronic structure of molecules. The theory requires a little further development to discuss more complicated electronic systems but it is based substantially on the same principles.

3.4 THE HYDROGEN MOLECULE

The classic discussion of the structure of the hydrogen molecule was made by Heitler and London in 1927. It is built upon the premise of perfect pairing of electrons in molecules. This is the quantum mechanical development of what had been the traditional way of thinking about the nature of the chemical bond. The whole basis of this approach, known more generally as the valence bond method, is to write down structures in which electrons are perfectly paired as the basis of the calculation.

Let the two electrons 1 and 2 move in the field of the two nuclei a and b which are a distance R_{ab} apart. Let the two electrons be at positions P_1 and P_2 as shown in the diagram below. The relevant interparticle distances are also shown on figure 7. The potential energy operator is

$$V = -\frac{1}{r_{a1}} - \frac{1}{r_{b1}} - \frac{1}{r_{a2}} - \frac{1}{r_{b2}} + \frac{1}{r_{12}} + \frac{1}{R_{ab}} \tag{18}$$

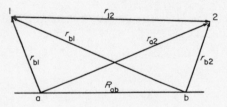

FIGURE 7.

The molecular orbital description would lead to placing the two electrons in the lowest molecular orbital. The total electronic wave function would then be a product of the function corresponding to each electron, and we have

$$\psi = \psi(1)\psi(2) = \{\phi_a(1) + \phi_b(1)\}\{\phi_a(2) + \phi_b(2)\}$$

$$= \{\phi_a(1)\phi_a(2) + \phi_a(1)\phi_b(2) + \phi_b(1)\phi_a(2) + \phi_b(1)\phi_b(2)\} \tag{19}$$

The first and last terms in (19) correspond to the states in which both electrons are located on one atom, leading to the structures H^-H^+ and H^+H^- respectively. Since the electron affinity of hydrogen is so much less than its ionization potential, these states will be unstable. Heitler and London suggested that a more physically realistic wave function is

$$\psi = \phi_a(1)\phi_b(2) + \phi_b(1)\phi_a(2) \tag{20}$$

in which there is perfect pairing in the ground state. The two ionic states are then rejected.

The energy corresponding to this wave function is given by

$$E = \frac{J' + K'}{1 + S^2} \tag{21}$$

where

$$J' = \int \phi_a(1)\phi_b(2)H\phi_a(1)\phi_b(2)\,d\tau_1\,d\tau_2$$

$$K' = \int \phi_a(1)\phi_b(2)H\phi_b(1)\phi_a(2)\,d\tau_1\,d\tau_2$$

$$S^2 = \int \phi_a(1)\phi_b(2)\phi_b(1)\phi_a(2)\,d\tau_1\,d\tau_2$$

From the form of the Hamiltonian and because ϕ_a and ϕ_b are hydrogen $1s$ wave functions (21) becomes

$$E = 2E_H + Q + \alpha \tag{22}$$

where

$$Q = \frac{J}{1+S^2} \quad \text{and} \quad \alpha = \frac{K}{1+S^2}$$

The integrals J and K are given by

$$J = \int \phi_a(1)\phi_b(2)\left(-\frac{1}{r_{b1}} - \frac{1}{r_{a2}} + \frac{1}{r_{12}} + \frac{1}{R_{ab}}\right)\phi_a(1)\phi_b(2)\,d\tau_1\,d\tau_2$$

$$= -\varepsilon_{aa} - \varepsilon_{bb} + \frac{1}{R_{ab}} + \int \phi_a(1)\phi_a(1)\phi_b(2)\phi_b(2)r_{12}^{-1}\,d\tau_1\,d\tau_2$$

where ε_{aa} is the integral defined in (16), and noting that $\varepsilon_{aa} = \varepsilon_{bb}$ we have

$$J = -2\varepsilon_{aa} + \frac{1}{R_{ab}} + \int \phi_a(1)\phi_a(1)\phi_b(2)\phi_b(2)r_{12}^{-1}\,d\tau_1\,d\tau_2 \tag{23}$$

Similarly it is seen that

$$K = \frac{S^2}{R_{ab}} - 2S\varepsilon_{ab} + \int \phi_a(1)\phi_b(2)\phi_a(2)\phi_b(1)r_{12}^{-1}\,d\tau_1\,d\tau_2 \tag{24}$$

where ε_{ab} is the other integral given in (16).

The only integrals left to evaluate are the two two-electron integrals, one in each of (23) and (24). Q and α are known as the Coulomb and exchange integrals. The Coulomb integral (which occurs in (23)) is the quantum-mechanical analogue of the Coulomb repulsion between two like electrostatic charges. The exchange energy has no classical analogue.

From equation (24), it can be seen that the integral K, which is substantially a measure of the bonding in the molecule, is roughly proportional to the overlap integral S, so that if the two orbitals overlap only slightly the exchange integral will be small and hence the binding energy will be small.

So far a discussion of spin has not entered into the argument. This is because the Hamiltonian is independent of electron spin, so that the orbital and spin wave functions are separable. As with electrons in atoms, we can write down two sets of spin wave functions for the two electrons. A singlet state given by

$$\frac{1}{\sqrt{2}}\{\alpha(1)\beta(2) - \alpha(2)\beta(1)\} \tag{25}$$

which is antisymmetric in the two electrons, and three functions which form a triplet given by

$$\alpha(1)\alpha(2)$$

$$\frac{1}{\sqrt{2}}\{\alpha(1)\beta(2) + \beta(1)\alpha(2)\} \tag{26}$$

$$\beta(1)\beta(2)$$

In order to satisfy the Pauli principle, the singlet antisymmetric spin state must be combined with a symmetric space function. Likewise the symmetric (triplet) spin state must be combined with an antisymmetric space function. So we have for the singlet the complete wave function (unnormalized)

$$\psi = \{\phi_a(1)\phi_b(2) + \phi_b(1)\phi_a(2)\}\{\alpha(1)\beta(2) - \beta(1)\alpha(2)\} \tag{27}$$

and for the triplet

$$\psi = \{\phi_a(1)\phi_b(2) - \phi_b(1)\phi_a(2)\} \begin{bmatrix} \alpha(1)\alpha(2) \\ \frac{1}{\sqrt{2}}\{\alpha(1)\beta(2) + \beta(1)\alpha(2) \\ \beta(1)\beta(2) \end{bmatrix} \tag{28}$$

The stable ground state of hydrogen is described by the singlet wave function (26).

Several simple modifications of the singlet state wave function can be introduced. For one thing, a variation parameter can be introduced, since there is no room for variation in the simple formulation of the problem. This can be achieved by using hydrogen-like wave functions

$$\phi_a(1) = \left(\frac{\alpha^3}{\pi}\right)^{\frac{1}{2}} \exp(-\alpha r_{a1}) \tag{29}$$

and similarly for ϕ_b. The quantity α is then taken as a variation parameter and the "best" energy determined by the variation theory criterion. Wang found the optimum value of α to be $1\cdot17$ and this gives a value for D_e of $3\cdot76$ eV, compared with the experimental value of $4\cdot72$ eV. It is this value which should be compared with the molecular orbital value. A more general form of the atomic wave function is obtained by allowing for the polarization of one orbital by the other. This will tend to modify the $1s$ orbital by introducing a certain amount of $2p$ character directed along the internuclear axis. This type of orbital introduces two variation parameters. Such a function is

$$\phi_{a1} = (1 + c_1 z_{a1}) \exp(-\alpha r_{a1}) \tag{30}$$

Rosen obtained values of $1\cdot17$ for α and $0\cdot10$ for c_1. This gives D_e a value of $4\cdot02$ eV.

A further modification is to introduce a certain amount of ionic character into the wave function. This can be achieved by including the terms omitted in (20) with a weighting factor c_2, regarding c_2 as a third variational parameter. The optimum energy is obtained for $\alpha = 1\cdot19$, $c_1 = 0\cdot07$ and $c_2 = 0\cdot175$, which gives a dissociation energy of $4\cdot10$ eV. This small value of c_2 justifies the Heitler–London assumption for the neglect of ionic terms (at least in the case of the hydrogen molecule).

When we go to other molecules, particularly heteronuclear molecules, the neglect of ionic terms leads to serious deficiencies.

3.5 COMPARISON OF THE THEORIES

It is useful at this point to examine the two theories and their performance. It has already been noticed that the molecular orbital theory overemphasizes the importance of ionic character. On the other hand the valence bond theory underestimates its significance. Whilst this is unimportant in the case of homopolar diatomics, in heteropolar diatomics the contribution of the ionic terms becomes important. The stage is reached when ionic terms acquire considerable importance in the valence bond function. At this point the basic premise of the theory no longer applies and it is questionable whether or not one can use the valence bond theory.

Similarly, the molecular orbital theory will always overemphasize the ionic character of the molecule in question, and its significance will always be questioned on this account.

It is instructive to consider the physical implication of the two theories. The molecular orbital theory is a logical extension of the orbital theory which has already been developed to discuss the electronic structure of atoms. The monocentric atomic orbital becomes quite simply the poly-centric molecular orbital. As the molecules increase to infinite dimensions the finite molecule becomes the crystal. The orbitals used to describe the electronic structure of the crystal are once again molecular orbitals which extend over the whole crystal, the polycentric molecular orbital having now become the delocalized orbital extending over all of the atoms in the macroscopic crystal. Herein lies the great importance of the orbital approach, for with simple extension and only slight modification, there is a theory which accounts for the electronic properties of atoms, molecules and crystals.

When we turn to the valence bond theory another set of advantages appear. No longer have we a comprehensive theory which accounts for the electronic structure of all states of matter, but the perfect pairing approximation gives a theoretical foundation for the chemical picture of the valence bond in a covalent molecule as it developed over the years. In cases where there are no alternative structures, the bonds are fixed in space as described by localized bond wave functions. These bonds are easily conceived and are of immense importance in a discussion of chemical properties. However, the number of molecules which can be usefully considered in this theory is severely limited. It is only by consider-able ingenuity that the theory can be extended to discuss the structure of covalent solids. It is totally inapplicable to the structure of atoms. In asses-sing the worth of the valence bond approximation, its chemical value must be balanced by its physical inadequacies as a comprehensive theory.

Returning to the hydrogen molecule, a simple molecular orbital ap-proach gives a value of 2.65 eV for D_e, as opposed to the value of 3.76 eV for the valence bond approximation. The experimental value is 4.72 eV. On the level of simple theory the valence bond method is obviously superior, but James and Coolidge, using a thirteen-term molecular orbital function, obtained a value of 4.698 eV. In 1959 Kolos and Roothaan, using a fifty-term variation function, obtained a value of 4.7466 eV which is obviously in excellent agreement with the experimental value. Similar values are obtained with the most sophisticated valence bond function. Here, when one is prepared to extend either theory to a very sophisticated—but no longer physically meaningful—basis, both approaches give good agreement.

3.6 HOMOPOLAR DIATOMICS

In what follows we shall use only the molecular orbital approach to molecular structure. The development of molecular orbital theory is due primarily to R. S. Mulliken, although of course advances of great significance have been made by other workers. At its simplest level, as we saw for H_2^+, the molecular orbital can be formed out of two atomic orbitals, one on each atom. If the atoms are identical, then the bond is "homopolar". In the case of a homopolar diatomic molecule, the molecular orbital will be given by

$$\psi = N^{-\frac{1}{2}}(\phi_a \pm \phi_b) \tag{31}$$

where ϕ_a and ϕ_b are the participating atomic orbitals and N is the normalization factor. The plus and minus combinations in (31) will be separated in energy (if bonding is to result), one being bonding (with respect to the electrons being localized on isolated atoms) and the other antibonding.

The molecular orbitals which result when two $1s$ orbitals combine have already been discussed. A rather similar state of affairs occurs if two atoms, both with a singly occupied p orbital, approach each other so that the p orbitals lie along the internuclear axis. In both cases molecular orbitals are formed which are cylindrically symmetric about the internuclear axis. Molecular orbitals which are symmetric about the internuclear axis are known as σ orbitals and are the molecular analogue of the atomic s orbitals (which are spherically symmetric).

However, if we form our molecular orbitals out of, say, $2p$ orbitals perpendicular to the internuclear axis, a different state of affairs occurs. The resultant molecular orbitals are no longer symmetric about the internuclear axis but are antisymmetric. They have a nodal plane which contains the internuclear axis. Molecular orbitals of this type are known as π orbitals, and are the analogue of the p orbitals in atomic orbital theory. The ways in which p orbitals can be combined to give molecular orbitals are shown diagrammatically in figure 8.

The molecular orbitals which result from combining p orbitals perpendicular to the internuclear axis are two-fold degenerate. $(p_{xa} + p_{xb})$ and $(p_{ya} + p_{yb})$ are degenerate. (We have used p_{xa} to represent ψ_{p_x} centred on atom a.) If there are two electrons available then by Hund's rule one electron will go into each molecular orbital. The electronic structure will then be $(\pi_x)^1(\pi_y)^1$ and the charge density will be

$$N(p_{xa}^2 + p_{xb}^2 + 2S_{ab}p_{xa}p_{xb} + p_{ya}^2 + p_{yb}^2 + 2S_{ab}p_{ya}p_{yb})$$

It is obvious that such a configuration will be symmetric about the internuclear axis. Any combination of these two π molecular orbitals will be an acceptable molecular orbital for the molecule. The two most significant orbitals are $\pi_\pm = (1/\sqrt{2})(\pi_x \pm i\pi_y)$, which turn out to be the π orbitals which are eigenfunctions of angular momentum about the internuclear axis.

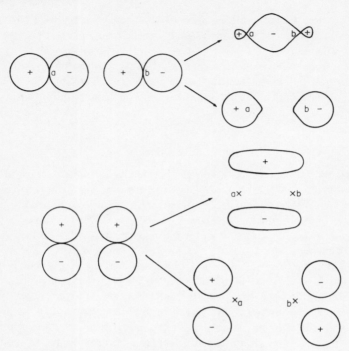

FIGURE 8. Molecular orbitals from *p* orbitals.

3.7 ANGULAR MOMENTUM

For the molecular state, the quantum numbers appropriate to the atomic case are no longer valid, with the exception of the spin quantum numbers. In the molecular case there is no longer a centre of symmetry and orbital angular momentum is no longer quantized. The quantum numbers l_z or m no longer have any significance. For linear molecules, however, there still remains one quantum number of angular momentum. This is the angular orbital momentum about the internuclear axis. The component of angular momentum about the internuclear axis is given the symbol λ.

Those molecular orbitals which are symmetric about the internuclear axis have λ values of zero. These are molecular orbitals formed out of *s* atomic orbitals or *p* orbitals whose lobes point along the internuclear axis and they correspond to the stomic *s* orbitals. Those molecular orbitals which change sign on reflection across a plane containing the internuclear axis are known as π orbitals and have a λ value of one;

these are the molecular analogue of the p atomic orbitals. Molecular orbitals with $\lambda = 2$ are known as δ orbitals and correspond to the d atomic orbitals.

Since most molecules contain more than one filled molecular orbital, the total angular momentum about the axis for the many-electron molecule is of considerable importance. The total angular momentum is given by

$$\Lambda = \sum_i \lambda_i \tag{32}$$

(all electrons)

Total electron states which are symmetric about the intermolecular axis are known as Σ, the analogue of the σ molecular orbital for the one-electron case. If only one electron is present and this occupies a σ orbital, the state of the molecule will be a Σ with $\Lambda = 0$. This is the case for the hydrogen molecule ion. Π states are those for which $\Lambda = 1$, and Δ those states for which $\Lambda = 2$.

Consider a molecule in which there are two electrons which can be allocated to π orbitals formed out of the two p orbitals on each atom perpendicular to the axis. From what was said above, the π orbitals, which are eigenfunctions of angular momentum, are not $(p_{x1} + p_{x2})$ and $(p_{y1} + p_{y2})$ but $\pi_x \pm i\pi_y$, where $\pi_x = (p_{x1} + p_{x2})$ and $\pi_y = (p_{y1} + p_{y2})$ and similarly for the combinations of the non-bonding pairs formed from $(p_{x1} - p_{x2})$ and $(p_{y1} - p_{y2})$. The two π_+ orbitals have an angular momentum of $+1$ about the internuclear axis whilst the two π_- orbitals have angular momenta of -1 unit about the axis. The state in which there are two electrons, one in each bonding π orbital, will have a total angular momentum of zero, whilst the state with four electrons will also be a Σ state. If we consider the half-filled case further, reflection across a plane containing the internuclear axis will interchange the electrons having angular momenta of plus and minus one unit, since the orbital with $+1$ unit will now become that with -1 unit and vice versa. This is equivalent to interchanging electrons and hence the wave function must change sign. Σ states are therefore characterized by a further symbol $+$ or $-$ added to the sigma as follows: Σ^+ or Σ^-. This indicates whether or not the wave function changes sign on reflection across a plane containing the internuclear axis. When the two orbitals are completely full, this operation corresponds to a double interchange and hence the sign of the wave function is unchanged. Molecules with completely filled sub-shells are therefore Σ^+ states.

Total electronic states are further characterized by adding the symbols g and u which indicate whether or not the wave function remains invariant or changes sign upon inversion through the centre of symmetry of the homopolar diatomic. It can be readily seen that the molecule which has one valence electron in an orbital $p_{x1} + p_{x2}$, changes sign upon inversion because the positive lobe of one orbital inverts to the place occupied by the negative lobe of the other orbital. Such a state is given the symbol u.

The homopolar diatomic molecule is characterized by three symbols. First a Greek capital letter, which indicates the total angular momentum about the internuclear axis; secondly (in the case of Σ states) a $+$ or $-$ added as a superscript which indicates whether or not the total electronic wave function changes sign upon reflection across a plane containing the internuclear axis (note that there is no one-electron analogue of this except when the molecule actually contains only one electron); and thirdly the symbol g or u is added as a subscript to indicate whether or not the wave function changes sign on inversion through the centre of symmetry of the molecule. (Note that this symbol can be applied to the one-electron case to indicate what happens to the individual orbital upon inversion.)

The total electron state is further characterized by the multiplicity of the state. This is determined from a consideration of the electron spin. The total electron spin eigenvalue is once again determined by adding together the s_z values for the individual electrons (in this case the inter-nuclear axis is the z axis). The total electron spin is

$$S = \sum_i s_i \tag{33}$$

Consider some examples:

1. *The Nitrogen Molecule N_2.* Each nitrogen atom has seven electrons and has the configuration $(1s)^2(2s)^2(2p)^3$. The $1s$ electrons on each atom do not participate in bonding. The four $2s$ electrons occupy both the bonding and antibonding orbitals formed out of the $2s$ atomic orbitals. This leaves six electrons which occupy the three molecular orbitals formed out of the three $2p$ atomic orbitals on each atom. One of these is a σ orbital and the remaining two are π orbitals. The antibonding orbitals are empty. The complete configuration of the molecule is

$$(1s)^2(1s)^2(\sigma 2s)^2(\sigma^*2s)^2(\sigma 2p)^2(\pi 2p_+)^2(\pi 2p_-)^2$$

These orbitals have angular momentum values of 0, 0, 0, 0, 0, $+1$, and -1 respectively. This leads to a total angular momentum

$$\Lambda = 2 \times 0 \pm 2 \times 0 + 2 \times 0 + 2 \times 0 \pm 2 \times 0 + 2 \times (+1) + 2 \times (-1) = 0$$

The ground state of the nitrogen molecule is a Σ state and because every orbital is doubly occupied it is also a singlet state. The wave function does not change sign upon reflection across a plane containing the N–N axis, and the state is also g, the ground state of the nitrogen molecule is given by the symbol $^1\Sigma_g^+$.

2. *The Oxygen Molecule O_2.* Each oxygen atom has one more electron than a nitrogen atom, and hence there will be two more electrons to accommodate in the ground state of the molecule. These two electrons will occupy the antibonding π_g^* orbitals. Since these are degenerate, following Hund's Rule one electron will go into each orbital, giving an electron configuration

$$(1s)^2(1s)^2(2s)^2(2s)^2(\sigma 2p)^2(\pi 2p_+)^2(\pi 2p_-)^2(\pi^* 2p_+)(\pi^* 2p_-)$$

The total angular momentum is zero leading to a Σ state. Reflection across the plane containing the O–O axis leads to a state which is equivalent to interchange of the two electrons in the antibonding π^* orbitals. It is therefore a $-$ state. The spins of the electrons in these two orbitals will be parallel, leading to a triplet ground state which has the term symbol $^3\Sigma_g^-$. From these considerations it is expected that the ground state of the oxygen molecule be paramagnetic and this is found to be the case.

3.8 HETEROPOLAR DIATOMICS

In our discussion so far we have been concerned with molecules in which both atoms are of the same type. When the atoms which make up the diatomic are different, the resultant heteropolar molecule has many features in common with the homopolar case and certain differences arise. In describing the electronic states of the many-electron heteropolar diatomic, the symbols u and g disappear since the molecule does not now possess a centre of inversion. The $+$ and $-$ signs which are used to characterize the Σ states are still applicable and are therefore used.

The most significant effect which occurs in a diatomic molecule with different atoms is that an unsymmetrical distribution of charge occurs. The molecular orbitals still have the form $c_A \phi_A + c_B \phi_B$ but $c_A \neq c_B$, and hence there is a non-uniform distribution of charge leading to a permanent dipole moment. The dipole moment is given by

$$\mu = \int \psi^* \mathbf{r} \psi \, d\tau \tag{34}$$

where **r** is the vector $x\mathbf{i} + y\mathbf{j} + z\mathbf{k}$. The components of the dipole moment are given by

$$\mu_x = \int \psi^* x \psi \, d\tau$$

$$\mu_y = \int \psi^* y \psi \, d\tau \tag{35}$$

$$\mu_z = \int \psi^* z \psi \, d\tau$$

The total dipole moment μ is given by

$$\mu^2 = \mu_x^2 + \mu_y^2 + \mu_z^2 = |\mu|^2 \tag{36}$$

Some typical values of dipole moments are given in the table below. The units are Debyes (one Debye $= 10^{-18}$ e.s.u. Å^{-1})

H—F	1·82 D	HBr	0·79D
HCl	1·07 D	H—I	0·38 D

In the valence-bond treatment of the hydrogen halides, it is found that ionic structures share equal importance with the perfect pairing structures for HF, and vary downwards from 5% as we go through from HCl to HI.

3.9   THE ENERGY OF THE DIATOMIC MOLECULE

Molecules with electronic closed shell configurations are represented by single Slater determinant wave functions in exactly the same way as atoms (see section 2.9).

The Hamiltonian for the N-electron molecule is (the nuclear motions being factored out)

$$h = -\frac{1}{2}\sum_i \nabla_i^2 + \sum_{iK} \frac{Z_K}{r_{iK}} + \frac{1}{2}\sum_{ij} \frac{1}{r_{ij}} + \frac{1}{2}\sum_{KL} \frac{Z_K Z_L}{R_{KL}} \tag{37}$$
$$\text{(all electrons)} \quad \text{(all nuclei)}$$

This equation differs from the atomic case [equation (73), Chapter 2] in the appearance of the nuclear repulsion term and in the appearance of the terms describing the interaction between the electrons and the adjacent nucleus (or other nuclei in the case of polyatomic molecules).

The formal analysis is exactly similar to that for the atomic case, and for the ground state, described by a single determinantal wave function, the electronic energy is given by

$$E = 2 \sum_i f_i + \sum_{ij} (2J_{ij} - K_{ij}) \tag{37}$$

(the subscripts here refer to orbitals).

To the energy given by (37) must be added the internuclear repulsion energy. The one-electron term now includes the attractive energy between an electron and the nucleus of the adjacent atom. The subscripts here refer to molecular orbitals and not to atomic orbitals. The evaluation of the integrals occurring in (37) gives rise to considerable difficulties. The two-electron integrals cannot be evaluated in closed form and various approximations have to be made to enable calculations to be made.

The orbitals which result as the solution of the eigenvalue problem associated with (37) are then used to calculate a new Hamiltonian and the process repeated until self-consistency is achieved. Many calculations along these lines have been carried through in recent years.

3.10 POTENTIAL ENERGY CURVES

The potential energy of a diatomic molecule is made up mainly of contributions from the vibration of the nuclei. To a first approximation the vibrating nuclei can be considered to be in simple harmonic motion. The energy of a vibrational state will therefore be given by

$$E_i = h\nu(n_i + \tfrac{1}{2}) \tag{39}$$

The quantum of vibrational energy is directly related to the electronic structure of the molecule through the force constant k, through the relation $\nu = (\tfrac{1}{2}\pi)(k/m)^{\tfrac{1}{2}}$ [cf. equation (56), Chapter 1].

If the vibrations of molecules were of the simple harmonic variety, then the plot of energy as a function of internuclear distance would be parabolic, as shown in figure 9.

However, whilst s.h.m. is a fairly good approximation for the ground state and the first few excited vibrations, marked deviations soon occur, leading to considerable modifications in the shape of the curve. The interspacing of adjacent levels is constant in the s.h.m. approximation, whilst in practice it is found that the levels are closer together at higher quantum numbers. In practice the potential energy curve of a diatomic molecule has the form as shown in figure 10.

If the molecule can be excited into the top quantum state, where the displacement is infinite, the molecule will fly apart. In other words, the

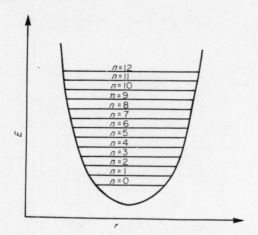

FIGURE 9. Energy levels for a s.h.o.

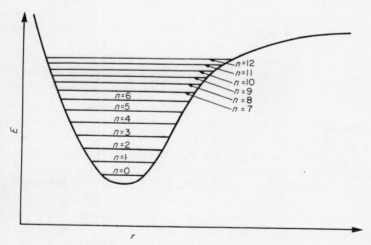

FIGURE 10. Energy levels for the vibration of a diatomic molecule.

energy required to put the molecule into this top quantum state is that energy which is required to dissociate the molecule. Spectroscopic measurements sometimes allow this energy to be measured but in the vast majority of cases it is necessary to estimate the convergence limit.

In the table on page 84, dissociation energies, equilibrium distances, vibrational quanta and force constants for the hydrogen halides are shown.

	D_e (kcal mole^{-1})	r_e (Å)	E (cm^{-1})	K (dynes cm^{-1})
HF	104	0·92	4141	$9·2 \times 10^{-5}$
HCl	102	1·27	2989	$5·1 \times 10^{-5}$
HBr	83	1·41	2650	$3·9 \times 10^{-5}$
HI	63	1·60	2309	$3·0 \times 10^{-5}$

It can be seen from this table that the strength of the bond decreases as we go down the group from fluorine to iodine. Also in accord with this the force constant decreases as the ionic character decreases.

3.11 EXCITED ELECTRONIC STATES

The transition to the excited state of a molecule can be visualized in terms of the promotion of a single electron from an orbital in the ground state to one of the orbitals which was unoccupied in the ground state. In fact, as we saw in the atomic case, this is a very naïve description, and only gives a general idea of what occurs. The excited state wave function is a linear combination of all configuration wave functions which possess the same symmetry as the state of interest. Whilst the single configuration describing the electron transfer under consideration may be the dominant term in the wave function, there will be contributions from other configurations. Secondly, it needs to be stated once more that the transition energy is not simply the difference between two orbital energies; there are additional terms to be considered involving two-electron integrals appropriate to the orbitals involved.

When a transition energy is calculated in this way, the assumption is made that the geometry of the excited state is identical to that of the ground state. In other words, the only thing which changes as the molecule becomes excited is that one of the electrons now occupies an orbital which was formerly empty.

This assumption is easy to justify theoretically. Because the electrons are so much lighter than the nuclei, electron motions are much quicker than those involving nuclei. Let us consider the transition of an electron from one orbital to another. The potential energy curves and vibrational quantum levels for the two states in question are shown in figure 11. The vibrational quantum levels are also indicated.

The lower curve represents the potential energy curve for the ground state. The upper curve represents the potential energy of the excited state

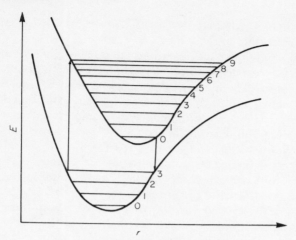

FIGURE 11. Transitions between two levels.

of interest. Its shape implies that it is of the same type as the ground state, i.e. a state which is stable and which does not immediately dissociate. It can be seen that this curve does not lie exactly above that for the ground state, but the position of the equilibrium (i.e. the internuclear separation) and its depth differs from that in the ground state. Also, the vibrational quantum is of a different magnitude (this is true in most cases). Since the excited state is less stable than the ground state its potential energy is higher.

To understand what occurs on excitation we have to use a principle known as the Franck–Condon principle. This principle states that because the nuclei are heavy compared with electrons, a change in electronic configuration occurs so quickly (of the order of 10^{-15} second) that the nuclei have no time to readjust until after the transition is completed. Since the internuclear separation does not change, it can be seen from the figure that an excited vibrational state results in the upper state. Such a transition is known as a "vertical" transition, and is the most intense band observed in the spectrum. The vibrationally-excited state can then lose energy by collision until it falls into the ground vibrational state of the excited electronic state. The Franck–Condon principle is just a principle and not a rigid physical fact, and so other transitions are observed which involve excitations to neighbouring vibrational levels, with decreasing intensity the further one moves away from the vertical transition. This means that the electronic spectrum of a diatomic molecule does not consist of a single line but a series of lines (each one of which is in fact a band of measurable width) corresponding to excitation to different vibrationally-excited levels of the upper electronic state.

In addition to all this other transitions are observed. These correspond to excitations from excited vibrational levels in the electronic ground state. These transitions will be much weaker since the population of the higher vibrational levels is proportional to the Boltzmann factor $\exp(-E/kT)$, and if E is greater than a few hundred wave numbers, the occupation of these higher levels will be insignificant.

In the diagram below, the intensity of the various transitions are plotted for a hypothetical diatomic molecule. It is important to notice the convention for writing the various transitions, $k-l$ implies the transition from lth vibrational level of the ground state to the kth vibrational level of the electronically excited state. In other words the convention is the opposite to what we would expect at first sight.

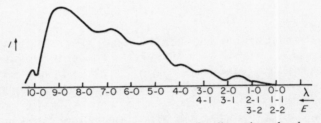

FIGURE 12. Absorption spectrum for diatomic molecule.

If we observe the band in emission it has a different appearance and occurs at a different frequency. This arises because emission occurs from the ground vibrational level of the excited state to the vibrational level of the ground state which lies vertically below it. The energy of this transition will be much lower than that of the vertical transition in absorption or excitation. If we plot the spectrum for emission on top of that for absorption, we find that the bulk of the emission spectrum lies on the low energy side of the absorption spectrum. This is shown below:

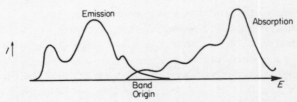

FIGURE 13. Emission and absorption for the same molecule.

The point at which the two curves overlap corresponds to the transition from the ground vibrational level to the ground vibrational level of the other electronic state. This point is known as the band origin and corresponds to the difference in energy between the vibrationally-unexcited electronic states.

When we look more closely at the vibrational "lines" in the electronic spectrum we find that each consists of a complicated series of lines which are very closely spaced (ranging from twenty to about a hundred wave numbers separation). The molecule is considered as a rigid rotor, and in the case of a diatomic molecule there are only two degrees of rotational freedom and these are degenerate. The energy levels of a rigid rotor are given by

$$E_J = \frac{h^2}{8\pi^2 I} J(J+1) \tag{40}$$

The possible transitions for a rigid rotor are $\Delta J = \pm 1$ and under certain circumstances in electronic–vibrational transitions $\Delta J = 0$. If $E_{v''}$ is the energy of the vibrational level from which the transition occurs and $E_{v'}$ is the energy of the level to which it goes in the excited electronic state, then the total energy involved if the rotational quantum number change is $+1, 0$ and -1 respectively is

$$J' = J'' + 1 \qquad E = E_{v'} - E_{v''} + \frac{2h^2}{8\pi^2 I}(J''+1)$$

$$J' = J'' \qquad E = E_{v'} - E_{v''} \tag{41}$$

$$J' = J'' - 1 \qquad E = E_{v'} - E_{v''} - 2J'' \frac{h^2}{8\pi^2 I}$$

The lines which correspond to $\Delta J = +1$ and have a spacing of $2(J''+1)h^2/8\pi^2 I$ are said to constitute the P branch of the spectrum, those with $J' = J''$ the Q branch and those with $\Delta J = -1$ and a spacing of $2J''h^2/8\pi^2 I$ are said to constitute the R branch. All molecules possess P and R branches and these are polarized parallel to the internuclear axis and perpendicular to the internuclear axis, whilst the Q branch is polarized perpendicular to the internuclear axis. The Q branch is therefore missing when the vibrational transition involved is polarized parallel to the internuclear axis.

As with vibrational lines, the separation between adjacent lines varies in the ground and excited electronic states. In the case of the rotational lines this is due to the fact that changes in molecular geometry give rise to a change in the moment of inertia in the upper state.

In the table below, some spectroscopic constants for a few molecules are given. Because of the Born–Oppenheimer approximation, it has been possible to discuss the contributions from purely electronic, vibrational and rotational effects independently of the others. It must be realized, however, that it *is* only an approximation. There is in fact some interaction between vibration and electronic states in many molecules which tends to upset the pattern already discussed. Fortunately the breakdown in the Born–Oppenheimer is second or third order and so only affects the discussion to a small extent.

	r_0 (Å)	D_e (eV)	$h\nu$ (cm^{-1})	l (gm cm$^2 \times 10^{-40}$)
Cl_2	1·989	2·481	564·9	114·8
CO	1·1284	9·144	216·8	14·48
HCl	1·275	4·431	2989	2·71
N_2	1·095	7·384	2360	13·94
O_2	1·2076	5·082	1580	1934

Normally, when we excite an electron in a molecule into a higher electronic state, the molecule reverts back to its normal ground state with emission of radiation after the exciting source has been removed, unless the molecule has been excited into a state which is unstable and in which

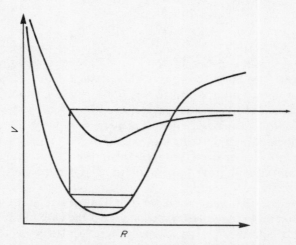

FIGURE 14. Energy curves for a diatomic molecule which predissociates.

the atoms fly apart. Thus if sufficient energy is added to excite the vibrational modes up to the onset of the continuum in the vibrational levels, dissociation will occur in the usual way. However, there is another possibility.

Consider the potential energy curves for the diatomic molecule shown in figure 14.

Here the stable part of the upper electronic state lies totally below the onset of the continuum for the ground state. The energy E' is the energy of some vibrational level of a stable molecule in the electronic ground state. In the excited electronic state, however, this energy level lies in the range of the continuous spectrum. In other words, in passing from the electronic ground state to the electronically-excited state the molecule automatically dissociates. This phenomenon is known as predissociation, and is a very important process which occurs in a wide range of molecules.

4
Small Molecules

4.1 INTRODUCTION

When we consider polyatomic molecules, there is nothing new introduced into the theory compared with the case of the diatomic molecule. In the molecular orbital theory the bicentric molecular orbitals are replaced by orbitals which now cover all of the nuclei. In the LCAO approximation, the orbitals have the form

$$\psi = \sum_i c_i \phi_i \tag{1}$$

Although these orbitals extend over the whole molecule, under certain conditions they approximate to localized bonds, and to a first approximation can be considered as such. These localized bonds have strong directional character.

On the other hand, a simple conceptual understanding can often be gained by using the valence-bond theory. Here, there is often the possibility of writing many purely covalent structures which interact to give the wave function. This mixing of structures leads to a lowering of energy and to stability of the molecule. In order to obtain any kind of quantitative agreement it is often necessary to include so many ionic structures that the fundamental assumptions of the theory are thereby threatened.

In what follows we shall examine the description of the structure of molecules using both the valence bond and the molecular orbital theories.

4.2 DIRECTED VALENCE

From what has been said previously, molecular orbitals formed through the combination of s atomic orbitals have no directional properties because the s orbitals are spherically symmetric in space. When we come to consider molecular orbitals formed out of p orbitals a new phenomenon appears. The p atomic orbitals are strongly directional in character. If we consider an atom with one electron in each of its p orbitals forming bonds with some other atom having a half filled atomic orbital, e.g. a hydrogen atom; then the linear combination of the three p orbitals which will minimize the electrostatic repulsion of the three hydrogen atoms occurs when these p orbitals are mutually at right angles to each other. This will tend, therefore, to result in a molecule in which the bonds are mutually at right angles to each other.

Consider the case of hydrogen sulphide. In its valence shell the sulphur atom has six electrons: two in the $3s$ orbital, two in one of the $3p$ orbitals and one in each of the other $3p$ orbitals. This leaves two p orbital vacancies on the sulphur atom. These enter into bonding with the $1s$ orbitals of the hydrogen atoms to give two localized bonds perpendicular to each other. The bond angle in the hydrogen sulphide molecule would therefore be expected to be 90°, and this is found to be so, the experimentally determined value being 92°.

Now consider carbon dioxide. The central carbon atom has four electrons in its valence shell, two in the $2s$ orbital and one in each of two of the p orbitals whilst the third p orbital is vacant. The two oxygen atoms each have six electrons in their valence shells, the $2s$ and one $2p$ orbitals are filled, whilst the other $2p$ orbitals each contain one electron. One possibility is that a bond is formed between one of the half-filled orbitals on the carbon atom and one of the half-filled p orbitals on the oxygen atom, giving two perpendicular bonds each formed out of two p orbitals. This leaves one unpaired electron on each oxygen atom whilst it leaves a completely unfilled p orbital on the carbon atom. A three-centred orbital would result in which the electrons would be donated by the oxygen atoms and in which the vacant $2p$ carbon orbital would participate. The resulting molecule would be bent and have a bond angle of 90°. In fact the carbon dioxide molecule is linear, in complete contradiction to what has been predicted. Some new factor has been introduced which changes the picture.

As a third example, consider the water molecule, which consists of a central oxygen atom and two hydrogen atoms. From what we said about the oxygen atom above, it would be expected that the water molecule would have a bond angle of 90°, exactly analogous to the hydrogen sulphide case. The bonds would be formed by overlapping the half-filled $2p$ orbitals with the $1s$ orbitals of the hydrogen atoms. The observed bond angle is 105°, which does not fit in with any form of bonding based on the principles already discussed.

As a last example, the ammonia molecule, following the above argument would be expected to have three $2p$–$1s$ bonds mutually perpendicular, having a bond angle of 90°, whilst the observed bond angle is 108°. The shape of the molecule would be a triangular pyramid, having the flat model

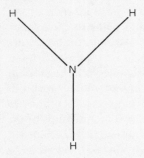

FIGURE 15.

which is in complete contradiction with experiment.

In all of the above cases we have made use of a criterion laid down many years ago by Linus Pauling that bonds are formed in the direction which leads to maximum overlap between the participating orbitals. Stable bonds result only when the overlap integral is large (in the case of hydrogen molecule it is of the order of 0·85). Pauling's second criterion requires that the energy difference between participating orbitals be not large.

When we considered the Heitler–London discussion of the hydrogen molecule, it was noted that to a first approximation the binding energy is proportional to the overlap integral, since the extent of bonding is determined by the exchange integral. This is the justification of the criterion laid down by Pauling.

When we come to consider s and p orbitals, we find that s orbitals overlap to a large extent, leading to strong bonds. The σ bond which is formed from the overlap of two p orbitals directed towards each other is weak, due to the fact that only a small part of each orbital is able to contribute to the overlap region because of the spatial distribution of the p_z orbitals. Such an overlap integral would certainly be of the order of 0·5 or less. When we come to consider the out-of-plane π orbitals formed by the overlap of p orbitals perpendicular to the internuclear axis, the overlap integral is of the order of 0·25, which one would expect to lead to a very weak bond.

In all of the cases discussed, since bonding is due to p orbitals on the central atom, we would expect all the molecules to be weakly bonded, whereas this is not the case. We must now examine the introduction of new factors which lead to strong bonding.

4.3 HYBRIDIZATION

In order to increase the strength of a bond, it is desirable to find a way of increasing the overlap integral between the participating orbitals. The way to do this can be understood in terms of one of the basic properties of atomic eigenfunctions (or indeed of any eigenfunctions). If ϕ_a and ϕ_b are two eigenfunctions of an operator h with eigenvalues E_a and E_b, the normalized sum of the eigenfunctions will also be an eigenfunction of the Hamiltonian with the eigenvalue $(E_a + E_b)N^{-1}$, where N is the normalization factor. This follows from the fact that the Hamiltonian is a linear operator. To find orbitals which overlap strongly it is necessary to find new combinations of the orbitals in the valence shell of an atom which are strongly directional and overlap to a large extent with orbitals on adjacent atoms. In order to do this we mix orbitals in the same valence shell, say s and p orbitals, first to give the p orbitals some s character which increases the strength of the resultant bonding, and secondly to give a set of strongly-directed bonds. In order to achieve this it is frequently necessary to use orbitals which would not otherwise be used, and to "prepare" the atom by putting it into a valence state.

Consider the case of carbon. The ground state configuration of the carbon atom is $(1s)^2(2s)^2(2p)^2$ and its lowest state is ^3P. If this configuration represents the state of the carbon atom when it forms bonds, then carbon would be divalent, forming two fairly weak bonds using p orbitals directed along the axis of the bond, each bond being at right angles to the other. In the example of "divalent" carbon which we considered above, i.e. the case of carbon dioxide, the molecule was linear, and not bent with a bond angle of 90°. If we look at the low-lying electronically-excited state of carbon, we find that by promoting one of the $2s$ electrons into the vacant $2p$ orbital, we could produce a ^5S (quintet S) state. This process requires an energy of about 4 eV. The ^5S state would be tetravalent and would give one strongly-bonded σ bond and three bonds formed out of the three p orbitals. The resultant state (in methane, for example) would lead to three identical bonds and one bond different from the others which would be strongly bonded. This is however contrary to the evidence, because in all molecules of the type CX_4, the four C—X bonds are identical. In order to obtain the correct state of affairs we *mix* the s orbital with the three p orbitals, and this leads to four orbitals which are strongly directional and which overlap very strongly with four orbitals on ligand atoms or groups. In order to obtain this state of the carbon atom we require more energy. Lowest estimates suggest about 0·5 eV whilst upper estimates suggest a value 3 or 4 eV higher. Stability is given to the system by the formation of these four extremely strong bonds which give rise to a molecule with greater binding energy than could have been obtained otherwise. All of this is summarized in figure 16.

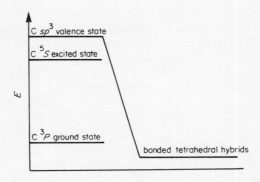

FIGURE 16. Energy of hybrid state and related states of carbon.

The state of the carbon atom from which the four bonds are formed is known as a valence state. It is important to note that this state is not a spectroscopic state and cannot be detected in an isolated carbon atom. If a methane molecule was simultaneously stripped of its four hydrogen atoms, the carbon would be observed to be in the ^5S state. Much confusion has arisen from a failure to realize that these valence states have no existence outside the bonding situation. It is absurd to talk of an atom isolated in space in its appropriate valence state.

Having discussed the valence state it is now necessary to see how the four functions are mixed in order to give the four orbitals which overlap strongly. We take the $2s$ and the three $2p$ orbitals and seek those four linear combinations which cause the overlap integral (with, say, a hydrogen $1s$ orbital) to be a maximum. We start with a variation function

$$\bar{\phi} = c_{2s}\phi_{2s} + c_{2px}\phi_{2px} + c_{2py}\phi_{2py} + c_{2pz}\phi_{2pz} \tag{2}$$

and seek the condition under which dS/dc_k is a maximum when carbon is bonded to hydrogen atoms. The four orbitals which result are

$$\bar{\phi}_1 = \tfrac{1}{2}(\phi_{2s} + \phi_{2px} + \phi_{2py} + \phi_{2pz})$$
$$\bar{\phi}_2 = \tfrac{1}{2}(\phi_{2s} + \phi_{2px} - \phi_{2py} - \phi_{2pz})$$
$$\bar{\phi}_3 = \tfrac{1}{2}(\phi_{2s} - \phi_{2px} + \phi_{2py} - \phi_{2pz})$$
$$\bar{\phi}_4 = \tfrac{1}{2}(\phi_{2s} - \phi_{2px} - \phi_{2py} + \phi_{2pz})$$

(3)

These orbitals are obtained by a variation calculation in which the overlap integral and not the energy has been optimized. The four orbitals which result as a consequence of the mixing of the four orbitals are strongly directional and point towards the vertices of a regular tetrahedron with the carbon nucleus as the centroid. Since these orbitals give maximum overlap with ligand orbitals, the exchange integral K must be larger than for the four unmixed orbitals, i.e.

$$|4K_h| > |K_{2s}| + 3|K_{2p}| \tag{3a}$$

where K_h is the exchange integral for the mixed orbital and K_{2s} and K_{2p} the exchange integrals for the $2s$ and $2p$ orbitals respectively. The increase in energy which results more than compensates the 4–7 eV required to promote the carbom atom from its ground state to its valence state.

This process of obtaining mixed orbitals from a set of atomic orbitals is known as hybridization and the resultant orbitals are known as hybrids.

The customary tetravalence of carbon is always achieved in this way, as the four bonds always point to the vertices of a regular tetrahedron. In the case of the molecule CWXYZ, where the central carbon atom is attached to four different atoms or groups, the bond angles may show a small deviation from 109°. This deviation occurs because the four attached groups will be different in size, will have different electrostatic repulsions and so will cause a slight deviation from the angle which occurs when all four ligand atoms or groups are identical.

Carbon attached to four other atoms or groups is not, however, the only environment in which we find the carbon atom. Even with hydrogen, carbon, as well as forming CH_4, also forms molecules of the type C_2H_4 (ethylene) in which the two carbon atoms are bonded together and the hydrogen atoms are bonded in pairs to the carbon atoms. In this environment the carbon atom appears to be trivalent, but the tetravalent state can be readily obtained from it by chemical reaction.

In a molecule such as ethylene there are obviously three bonds of σ character in association with each carbon atom and a further bond between the carbon atoms, much weaker and of a different nature. This state of affairs can be understood in the following way. The carbon atom is raised to the quintet S state in which one electron is found in the $2s$ orbital and in each of the $2p$ orbitals. The atom is then considered to be converted into its valence state,.in which the s electron and two of the p electrons are assigned to 3 hybrid orbitals formed out of the s orbital and two of the p orbitals having maximum overlap with ligand orbitals. These hybrid orbitals are

$$\bar{\phi}_1 = \frac{1}{\sqrt{3}}\phi_{2s} + \sqrt{\frac{2}{3}}\phi_{2px}$$

$$\bar{\phi}_2 = \frac{1}{\sqrt{3}}\phi_{2s} - \frac{1}{\sqrt{6}}\phi_{2px} + \frac{1}{\sqrt{2}}\phi_{2py} \tag{4}$$

$$\bar{\phi}_3 = \frac{1}{\sqrt{3}}\phi_{2s} - \frac{1}{\sqrt{6}}\phi_{2px} - \frac{1}{\sqrt{2}}\phi_{2py}$$

These three hybrid orbitals are coplanar and are directed at an angle of 120° to each other. There remains one p orbital which is perpendicular to the plane of the hybrid orbitals (often referred to as sp^2 hybrids). This is a p orbital which is able to combine with another p orbital in an adjacent carbon atom (or in some cases nitrogen or oxygen) to form a weak π bond. The structure of ethylene consists of one σ bond linking the carbon atoms and four other coplanar bonds linking the four hydrogen atoms to the two carbon atoms, together with one π bond between the two carbon atoms. This picture accounts for the fact that it is easy to break one of the bonds (the π bond) to form addition compounds with ethylene, for the π bond will be weak.

The possible bonding methods of compounds containing carbon and hydrogen are not yet exhausted. There is a further class of compounds in which the carbon appears to be divalent, in the sense that it is attached to only two other atoms. The simplest case is acetylene, C_2H_2, which although it is a stable molecule can, nevertheless, add two further mono-valent atoms or groups to each carbon atom. Since the molecule is not a free radical, the only explanation is that there is one strong bond linking together the two carbon atoms and a further two π bonds binding together the carbon atoms and the two hydrogen atoms.

The hybridization of the carbon orbitals which leads to the formation of these bonds can be understood as follows. From the ^5S state of the carbon atom, the valence state in which one s orbital and one p orbital are mixed is formed. The two remaining orbitals, each containing one electron, are p orbitals to be used subsequently in π bonding, and play no part in the hybridization. The hybrid orbitals are given by

$$\bar{\phi}_1 = \frac{1}{\sqrt{2}}(\phi_{2s} + \phi_{2px})$$

$$\bar{\phi}_2 = \frac{1}{\sqrt{2}}(\phi_{2s} - \phi_{2px})$$

(5)

These hybrid orbitals are usually known as sp hybrids and point in opposite directions to each other, having a bond angle of 180°.

In the acetylene molecule the carbon–carbon bond is a σ bond formed by the overlap of one hybrid orbital on each carbon atom. The remaining sp hybrid on each carbon atom is then used to form a σ bond with a hydrogen atom. The remaining electrons in the p orbitals then combine to give two π orbitals. It is important to realize that these doubly-filled π molecular orbitals lead to an electron distribution which is cylindrically symmetric about the internuclear axis.

The sp^3, sp^2 and sp hybrid orbitals are usually referred to as tetrahedral, trigonal and digonal hybrids respectively. From the form of hybrid orbitals it is possible to calculate the percentage s character in each type of hybrid, and this is found to be 25% for sp^3, 33% for sp^2 and 50% for sp hybrids. This should be reflected in the bond characteristics for the three types of bond, and in fact the force constants (which we noted in the last chapter are an indication of the tightness of the bonding) have the values 4.50×10^{-5} dyne cm^{-1} for C—C, 9.6 for C=C and 15.59 for C≡C.

Since a π bond is much weaker than a σ bond it is expected that an ethylenic carbon–carbon bond will have less than twice the bond energy of a carbon–carbon single bond (as found in fully saturated hydrocarbons); similarly for the acetylenic bond it should be less than three times that for the single bond. The observed experimental values of the bond strengths are C—C 2.56 eV, C=C 4.3 eV and 5.35 for C≡C, in good agreement with the predicted behaviour.

This discussion of the hybridization in the carbon atom allows us to understand the structure of the carbon dioxide molecule. It will be remembered that in the last section it was predicted that the carbon dioxide would be bent to a bond angle of 90° in accord with the bonding being due to p orbitals. In fact, as was mentioned before, the molecule is found to be linear. The carbon atom has two sp hybrids which are used to form σ bonds with orbitals on the oxygen atoms. The remaining two electrons on the carbon atom form π bonds with the electrons in the half-filled p orbitals on the oxygen atoms. This description accounts in a qualitative way for the behaviour of carbon dioxide and for the fact that

Delaware Valley
College Library

the molecule is linear in its ground state. In a more sophisticated approach it is found that the σ bonding is in fact a little more complicated than we have discussed: the lowest-lying σ orbital is tricentric and covers the whole molecule. However, the main details are explained using this model for the electronic structure.

The other molecules discussed above were water and ammonia. Here another factor is introduced. In what was discussed above, the sole criterion was to construct hybrid orbitals which maximized the overlap region between orbitals. In molecules where there are no lone pairs this is a good criterion, but where there are we must take into account the electrostatic repulsion between these localized charges. This leads to a partial hybridization of, say, the bonding orbitals in the water molecule and a different state of affairs in the hybridization of the lone pairs. This accounts for the fact that the bond angle in water is not 90°, as one would expect for pure p bonding, nor is it the 109° that one would expect for an sp^3 hybrid. Instead it is somewhere between the two, in fact 105°. This degree of sp hybridization, accompanied by the appropriate hybridization of the lone pairs, strikes the compromise between increasing the bond energy due to the increased s character in the bonds and minimizing the repulsion between the two large areas of negative charge in the lone pairs on the oxygen atom.

In the case of the ammonia molecule there is only one lone pair. As we noted above the structure of this molecule is a triangular pyramid, the bond angle being 108°. This is very close to the ideal angle for sp^3 hybridization. We notice how readily the ammonia molecule attaches a proton to itself to give the ammonium ion NH_4^+, in which real tetrahedral bonding occurs as in methane.

When we come to include d orbitals in the scheme then a whole new range of spatial possibilities opens up.

4.4   VALENCE BOND THEORY

In discussing the use of valence bond theory in the theory of polyatomic molecules several facts need to be noted. The number of structures which contribute to the wave function of the molecule are usually several in number. Except in the simplest homopolar molecules it is impossible to ignore the contribution of ionic structures to the wave function.

When writing down the possible structures which occur in the wave function we must note the consequences of the Born–Oppenheimer approximation. This requires that the only structures which are involved in the wave function are those which have the same spatial distribution of the nuclei. Valence bond structures differ only in their assignment of the electrons in the molecule. Further, we note that the wave function is written as

$$\Psi = \sum_l a_l \Phi_l \qquad (6)$$

where the Φ_l are the wave functions corresponding to the individual structures. The a_l are then determined by a variation calculation. The contributing structures to the valence bond wave function have frequently been called canonical structures, because they represent certain extreme possibilities for the structure of the molecule, and these will be in terms of localized single and double bonds and also completely ionic structures. In fact the true structures lies somewhere between them all. It is important to note however that these individual structures have no physical existence. The entities which could exist physically and are the analogues of the resonating structures would differ radically in their molecular geometries.

In the cases discussed above, the valence bond theory sheds some light. The valence bond theory makes use of hybrid orbitals in exactly the same way as the molecular orbital theory.

Carbon dioxide can be represented by the single valence bond structure

$$O=C=O$$

However, in addition to the purely homopolar structure it is necessary to include four ionic structures:

$$O=C^+-O^- \qquad O^--C^+=O$$
$$O^+-C-O^- \qquad O^--C-O^+$$

The lower two structures play an extremely important role, for they cause a raising of the bond energy and also a reduction of the interatomic distances. These ionic structures play an exactly similar role in the structure of carbon suboxide, C_3O_2.

Turning to ammonia, the role of the ionic structures is even more marked. In addition to the covalent structure

$$\begin{array}{c} H \\ | \\ N \\ \diagup \quad \diagdown \\ H \qquad H \end{array}$$

there are seven ionic structures which must be included. These are:

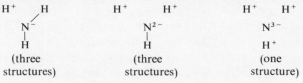

$H^+ \quad H$	$H^+ \quad H^+$	$H^+ \quad H^+$
N^-	N^{2-}	N^{3-}
H	H	H^+
(three structures)	(three structures)	(one structure)

The first is the most important contributor to the wave function for the ground state but the others cannot be ignored.

Compounds of the ammonium type ($NH_4^+X^-$) are very stable, because in the process of conversion of nitrogen to the tetravalent state, an additional covalent bond is formed. The energy of the molecule is increased by the difference between the electron affinity of the halogen X (plus the Coulombic energy of the ionic bond between the ammonium ion and the halogen ion) and the ionization potential of ammonia. There is a close parallel here between the compounds of nitrogen and the compounds of boron. Ammonia achieves tetravalency by forming four covalent bonds and losing an electron to some outside atom or group, whilst boron achieves the same end by obtaining an additional electron from some outside atom or group in order to form four covalent bonds. The analogous boron molecules will therefore involve ionic species in which the boron part is the cation, e.g. BF_4^-.

In water there is the covalent structure

$$\underset{H \qquad\quad H}{\overset{O}{\diagup \quad \diagdown}}$$

but also included in the ground state wave function there are contributions from the structures

$$\underset{H^+ \qquad H}{\overset{O^-}{\diagdown}} \qquad\qquad \underset{H^+ \qquad H^+}{\overset{O^{2-}}{}}$$

(two structures)

The most significant contribution comes from the two singly-ionized structures, the contribution of the doubly-charged structure being slight.

4.5 VIBRATIONS OF POLYATOMIC MOLECULES

When we consider the vibrations of polyatomic molecules, the situation is much more complicated than in the case of a simple diatomic molecule. For a diatomic molecule there is only one vibration possible: the stretching of the bond joining the two atoms. In the polyatomic molecule there are many more possible vibrations. In general, if there are n atoms in the molecule there will be $3n-6$ independent vibrations. This number occurs because the total number of degrees of freedom is $3n$ and three of these are taken up in translational motion and a further three in rotational motion, the remainder being vibrational degrees of freedom. In the case of linear molecules there are only two degrees of rotational freedom and therefore there are $3n-5$ degrees of vibrational freedom.

There are two types of vibration in polyatomic molecules. The first involves the stretching of bonds. This is similar to the vibrations in diatomics and usually requires energies of the order of 1000–2000 cm^{-1}. The other type of vibration is the bending of the molecule without changing the internuclear separation of nearest neighbours (or bonded atoms). This is a much easier process and requires less energy. The bending vibrational quanta have energies usually of the order of a few hundred wave numbers.

It is now necessary to examine the basis of vibrational theory a little further, and to see what happens when we consider a vibrating molecule in which the nuclei act as point masses.

Consider a set of vibrating particles, each only being disturbed slightly from its equilibrium position. Let us choose a set of coordinates such that all of the coordinates q_i vanish at equilibrium; then all the q_i, which represent displacements, will be small.

In Cartesian coordinates, the kinetic energy of the set of particles is

$$T = \frac{1}{2} \sum_i m_i \left(\frac{dx_i}{dt} \right)^2 \tag{7}$$

In general, using a set of coordinates q_i we have

$$T = \frac{1}{2} \sum_{ij} a_{ij} \frac{dq_i}{dt} \frac{dq_j}{dt} \tag{8}$$

To a reasonable degree of accuracy we can regard the a_{ij} as constant.

The potential energy of the set of particles can be expanded as a Taylor's series in the q_i about the point of equilibrium and we obtain

$$V = V_0 + \sum_i \left(\frac{\partial V_i}{\partial q_i} \right) q_i + \sum_{ij} \frac{1}{2} \left(\frac{\partial^2 V}{\partial q_i \, \partial q_j} \right) q_i q_j \tag{9}$$

where the differential coefficients are evaluated at the points $q_i = 0$. The quantity V_0 may be assumed to be zero, since this term merely raises or lowers the total energy by a constant amount. Since the state in which the q_i are all zero is the equilibrium configuration, the potential energy must be a minimum at this point, therefore all of the $(\partial V / \partial q_i)$ will be zero. This leaves only the last term in (9) non-vanishing. Writing b_{ij} for $\partial^2 V / (\partial q_i \, \partial q_j)$ we have

$$V = \frac{1}{2} \sum_{ij} b_{ij} q_i q_j \tag{10}$$

In order to proceed further we must put the equations for the motion of the oscillating particles into the Lagrangian form, for which

$$\frac{d}{dt} \frac{\partial T}{\partial q_j} - \frac{\partial}{\partial q_j} (T - V) = 0 \tag{11}$$

leading to the i equations

$$\sum_j a_{ij}\frac{d^2q_j}{dt^2} + \sum_j b_{ij}q_j = 0 \tag{12}$$

If the system has F degrees of freedom, then there will be F of these equations. To solve the set, we choose some constants c_i, such that if each equation is multiplied by one of the set and they are all added together we obtain

$$\frac{d^2Q}{dt^2} + \lambda Q = 0 \tag{13}$$

where

$$Q = \sum_j h_j q_j$$

In other words we seek the transformation which simultaneously diagonalizes the kinetic and potential energy matrices. We wish to obtain

$$\sum_i c_i a_{ij} = \frac{1}{\lambda}\sum_i c_i b_{ij} = h_j \tag{14}$$

In order to do this we write

$$\sum_i (\lambda a_{ij} - b_{ij})c_i = 0 \tag{15}$$

The only non-trivial solution occurs when

$$|\lambda a_{ij} - b_{ij}| = 0 \tag{16}$$

Having found the value of λ which are the eigenvalues of this determinant, we determine the eigenvectors c_i, from which the h_j's are now fixed. The h's being known, the coordinates Q can be easily determined. These coordinates Q are known as normal coordinates. In terms of the normal coordinates we have

$$T = \frac{1}{2}\sum_i \left(\frac{\partial Q_i}{\partial t}\right)^2 \qquad V = \frac{1}{2}\sum_i \lambda_i Q_i^2 \tag{17}$$

If the equilibrium is stable, all of the $\lambda_i(=\partial^2 V/\partial Q_i^2)$ are real and positive. For positive λ

$$Q_i = A_i \cos(\lambda_i^{\frac{1}{2}} t + \varepsilon_i) \tag{18}$$

where A_i and ε_i are arbitrary constants.

In terms of our original coordinates we have

$$q_i = \sum_j g_{ij} Q_j \tag{19}$$

$$= \sum_j g_{ij} A_j \cos (\lambda_j^{\frac{1}{2}} t + \varepsilon_j) \tag{20}$$

When all of the A_j are zero except one, each q_i will vary sinusoidially with time and they will all be in phase. This sort of motion is called a normal mode. Corresponding to each normal mode there is a frequency

$$v_j = \frac{\sqrt{\lambda_j}}{2\pi} \tag{21}$$

When vibrations are observed, what is seen is a superimposition of the possible normal modes of the system under observation, and these normal modes will have arbitrary amplitudes and phases.

If the equilibrium is unstable, the above treatment is still valid, except that one or more of the eigenvalues λ_i will be real and negative, and therefore at least one v_i is imaginary. The motion will not then be sinusoidal, since the corresponding Q will be an exponentially increasing function of time.

Returning now to the molecular case, and treating the molecule as a system of point masses (which are the nuclei), the kinetic and potential energies are given by

$$T = \frac{1}{2} \sum_{k=1}^{3N} \left(\frac{\partial Q_k}{\partial t} \right)^2 \qquad V = \frac{1}{2} \sum_{k=1}^{3N} \lambda_k Q_k^2 \tag{22}$$

where N is the total number of atoms.

Choosing any arbitrary coordinate system we may write

$$T = \frac{1}{2} \sum_{ij} a_{ij} \frac{dQ_i'}{dt} \frac{dQ_j'}{dt} \qquad V = \frac{1}{2} \sum_{ij} b_{ij} Q_i' Q_j' \tag{23}$$

The values of λ are then obtained from the determinant

$$|a_{ij}\lambda - b_{ij}| = 0 \tag{24}$$

Any set of coordinates Q' may be chosen, and it will be a coincidence if these happen to be normal coordinates. They may be chosen as atom or bond displacements, bond angle changes or any other meaningful set, for provided that the correct number of coordinates are used, the result will be correct. If the Q' are chosen to reflect the symmetry properties of the molecule, then considerable simplification of the problem occurs, for cross terms in (23) which have Q''s with different symmetry properties will vanish. This set of coordinates can be expressed as a linear sum of the normal coordinates for the molecule. In terms of the normal coordinates the vibrational wave function is

$$\sum_{i=1}^{3N} \frac{\partial^2 \psi}{\partial Q_i^2} + \frac{8\pi^2 m}{h^2}\left(E - \frac{1}{2}\sum_{i=1}^{3N} \lambda_i Q_i^2\right)\psi = 0 \tag{25}$$

We assume the wave function to be of the form

$$\psi = \prod_{i=1}^{3N} \psi_i(Q_i) \tag{26}$$

using the usual technique of separation of variables, which leads to

$$E = \sum_{i=1}^{3N} E_i \tag{27}$$

for the total vibrational energy of the system. Substituting (26) and (27) into (25) we obtain the $3N$ equations

$$\frac{\mathrm{d}^2\psi}{\mathrm{d}Q_i^2} + \frac{8\pi^2 m}{h^2}\left(E_i - \frac{1}{2}\lambda_i Q_i^2\right)\psi_i = 0 \tag{28}$$

Each of these equations is the equation for a simple harmonic oscillator whose eigenvalues and eigenvectors are already known.

The total wave function (26) will be of the form

$$\psi_{n1,n2,\ldots} = \exp\left(-\frac{1}{2}\sum_i \alpha_i Q_i^2\right) \prod_i H_{ni}(\sqrt{\alpha_i}Q_i) \tag{29}$$

where H_{ni} is the Hermite polynomial of degree ni in $(\sqrt{\alpha_i}Q_i)$.

A transition is allowed only if the transition dipole moment has at least one non-vanishing component. Considering the transition $\psi_{n1,n2,\ldots,n_{3N}} \to \psi_{n1',n2',\ldots,n'_{3N}}$, this is allowed only if

$$\int \psi^*_{n1,n2,\ldots n_{3N}} x \psi_{n1',n2',\ldots,n'_{3N}}\,\mathrm{d}\tau \neq 0$$

or   (30)

$$\int \psi^* y\psi'\,\mathrm{d}\tau \neq 0, \qquad \int \psi^* z\psi'\,\mathrm{d}\tau \neq 0$$

Interest centres on the transition in which the excited state differs from the ground state in the coordinates of one vibration, i.e. a fundamental, say the state in which ψ_{0i} is excited to ψ_{1i}. This transition will be infrared active only if $\psi_{0i}\psi_{1i}$ transforms like x, y or z. Since the ground state is symmetric, the only way in which a nonvanishing integral in (30) can arise is if the excited state transforms like x, y or z. In the case of a fundamental this implies that $\Delta n = \pm 1$.

Consider the vibrations of the water molecule. It can be shown from group theoretical considerations that the possible vibrations are three in number, two of which are completely symmetric with respect to rotation about the two-fold symmetry axis and to reflection across the plane containing the symmetry axis, and the other which changes sign on reflection and rotation. In other words two do not change sign when the hydrogen atoms are interchanged whilst the remaining one does change sign when the atoms are interchanged.

To find the actual normal coordinates it is necessary to solve the secular equation using some assumed form for the potential energy. We can however do a great deal by examining the symmetry properties. Consider that the mass of the oxygen atom is infinite compared with the mass of the hydrogen atoms; in other words we consider the oxygen atom not to move in the course of the vibration. Then we can write down symmetry coordinates

$$Q_1' = a'\phi + b'(r_1 + r_2)$$
$$Q_2' = c'\phi + d'(r_1 + r_2) \tag{31}$$
$$Q_3' = e'(r_1 - r_2)$$

where ϕ is the change in the bond angle and r_1 and r_2 are the extensions of the two bonds. Since Q_3' differs in symmetry from the other two co-ordinates, the cross terms $Q_1'Q_3'$ and $Q_2'Q_3'$ will vanish in the secular determinant. Solution of the secular determinant shows that one of the symmetric vibrations represents the bending of the molecule whilst the other is involved with the stretching of the two O—H bonds. The normal coordinates are thus

$$Q_1 = a\phi \qquad Q_2 = b(r_1 + r_2) \qquad Q_3 = c(r_1 - r_2) \tag{32}$$

These three vibrations can be shown diagrammatically:

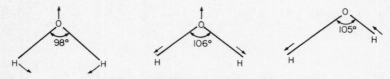

FIGURE 17. Vibrations of the water molecule.

Solution of the determinant must be carried out to evaluate the force constants and this requires some assumption about the potential energy.

If it is assumed that the z axis is the symmetry axis, then the totally symmetric vibrations transform like z, and both of these vibrations are allowed, the antisymmetric vibration transforms like x, and therefore all three appear in the infrared.

It is found from experimental data that the frequencies for the normal modes in water are 3652 cm^{-1} for the symmetric stretching, 3756 cm^{-1} for the antisymmetric stretching and 1545 cm^{-1} for the bending mode. It can be seen that the bond bending is much easier energetically than either of the possible bond stretching vibrations.

The formulation of the potential energy has always given rise to considerable difficulties. There have been two straightforward attempts to define a realistic force field. The first is based on interatomic attractions and repulsions and is known as the central force field, whilst the second attempts to express the forces in terms of forces directed along valence bonds and associated with the bending of valence angles. Such a force field is known as a valence force field.

In the central force field, the potential energy is expressed as a sum of the attractions and repulsions between all pairs of atoms.

$$V = \sum_i a_{ii} Q_i'^2 \tag{33}$$

for all coordinate changes, and in the non-linear XY_2 case, of which the molecule H_2O is an example, we have

$$2V = a_{11}(Q_1^2 + Q_2^2) + a_{33}Q_3^2$$

where Q_1' and Q_2' are the extensions in the two O—H bond lengths and the (force constant $\times Q_i'$) represents the energy between the oxygen atom and one of the hydrogen atoms: $a_{33}Q_3'^2$ represents the potential energy between the two hydrogen atoms.

For a few selected molecules we have the following data, where the force constants have been calculated on the basis of the central force

	v_1 (cm^{-1})	v_2 (cm^{-1})	v_3 (cm^{-1})	a_{11} ($\times 10^5$ dynes)	a_{33} ($\times 10^5$ dynes)
H_2O	3652	1595	3756	7·76	1·85
H_2S	2611	1290	2684	4·14	0·95
SO_2	1151	524	1362	9·97	3·24
NO_2	1320	648	1621	9·13	4·34

model. The force constant a_{11} refers to that between the oxygen and a hydrogen atom, whilst a_{22} refers to that between the two hydrogen atoms in the water case, and the analogues in the other cases.

If we now consider the valence force approximation, the potential energy is represented by

$$2V = \sum_i k_i Q_i'^2 + \sum_j k_{\delta_j}\delta_j \tag{34}$$

where the Q_i' represents the stretching of the valence bonds and δ_j represents the displacement in the bond angles.

For the non-linear XY_2 molecules discussed above we obtain the following force constants

	k_i ($\times 10^5$ dynes)	k_δ ($\times 10^5$ dynes)
H_2O	7·76	0·69
H_2S	4·14	0·45
SO_2	9·97	0·81
NO_2	9·13	1·52

In fact, neither of these force fields is satisfactory as a general method and more complicated force fields have been used. For example, in the force field formulated by Urey and Bradley, the valence force field is used, with the addition of terms which represent the forces between non-bonded atoms from the central force field approximation.

From what has been said above [equation (30), Chapter 4] about the intensities of vibrational transitions in the infrared spectrum, it is noted that a vibration is allowed only if there is a transition dipole moment change during the transition. In the case of homopolar diatomics this is impossible and there is no pure vibrational spectrum in these cases. The vibrations become allowed in the electronic spectrum for there is a change in dipole moment in the excited electronic state.

In some polyatomic molecules in which there is no permanent dipole moment, a dipole moment is produced in the course of a vibration and hence the transition is allowed. A molecule of this sort is carbon dioxide. The possible vibrations for the carbon dioxide molecule are shown in figure 18.

The vibration (a) is symmetric and there is no transition dipole moment developed in the course of the vibration, so this vibration will not occur in the infrared spectrum. The stretching vibration (b) is not symmetric

Stretching
vibrations

Bending
vibration

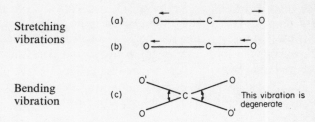

FIGURE 18. Vibrations of carbon dioxide.

and there will be a dipole moment set up during the course of the vibration; similarly the degenerate bending vibration (c) causes the centres of positive and negative charge to be different, and there will be a dipole moment change. Two of the three vibrations will thus be active in the infrared spectrum. This is in fact found—there are two intense bands with energies of 2349 and $667 \, \text{cm}^{-1}$. The other vibration cannot be detected in the infrared spectrum.

Vibrations which cannot be detected in the direct way can often be seen in the Raman spectrum. In the Raman method, a specimen is subjected to some exciting radiation, often from a mercury lamp. The molecules are excited in their vibrational modes and a beam of scattered light is observed at right angles to the incident beam. We shall not discuss the theory of Raman scattering here, except to note that the intensity of a Raman line does not depend on a change in dipole moment but upon a change in polarizability on vibration. It is therefore often possible to observe lines which are forbidden in the infrared. The vibration of carbon dioxide which is forbidden in the infrared appears in the Raman spectrum and has an energy of $1340 \, \text{cm}^{-1}$. The other two carbon dioxide frequencies are not Raman active. This indicates another important rule in vibrational spectroscopy: molecules which contain a centre of inversion never have vibrations active in both the infrared and the Raman. The fact that carbon dioxide has two vibrations active in the infrared and the other in the Raman is proof that the carbon dioxide molecule is linear and possesses a centre of inversion. On the other hand, all the water vibrations are active in both the infrared and Raman. The water molecule therefore contains no centre of inversion, and hence is not linear but bent; this we know to be the case.

Experimentally, Raman spectroscopy has been very difficult because until recently it has proved impossible to obtain a high intensity incident light source. Use of laser beams has now solved this problem, and this should lead in turn to a much more extended use of the Raman method.

A complete analysis of the vibrations of a molecule is possible only for small molecules or for larger molecules with high symmetries. For large asymmetric molecules the number of vibrational degrees of freedom becomes very large and an exact assignment is impossible in most cases. In the infrared spectrum of such molecules, certain frequencies occur associated with the presence of particular groupings in the molecule. These characteristic frequencies are typical either of a particular grouping of atoms or else of a particular type of bond. They appear to be independent of whatever else the molecule may contain. This can be illustrated by examining the energy of the $C{=}O$ stretching vibration in the molecule R_1COR_2 where the groups R_1 and R_2 are widely different in character.

| R_1 | | | R_2 | | |
	NH_2	C_6H_5	H	OC_2H_5	Cl
NH_2	1655	1652	1671	1692	1731
C_6H_5	–	1653	1696	1715	1768
H	–	–	–	1715	–
OC_2H_5	–	–	–	1743	1772
Cl	–	–	–	–	1810

In acetone, $(CH_3)_2CO$, the energy is $1706 \, cm^{-1}$. In ketones and aldehydes the variation lies between 1700 and $1725 \, cm^{-1}$. Considering the very wide nature of choice of the groups in the table it is surprising that the spread is only $150 \, cm^{-1}$, or less than 10%. All of the cases which lie on the low side of $1700 \, cm^{-1}$ can be understood in terms of conjugation with the groups R_1 and/or R_2, or else, in valence bond terminology, resonance between several possible structures.

The vibrational quanta for several multiple bonds are given in the table on page 110.

The analysis of the infrared spectrum of a large molecule (whose structure is either totally or partially undetermined) can sometimes be aided considerably by an examination of the vibrational spectra both in the infrared and in the Raman. It is also possible to make a reasonable quantitative estimate of the extent to which a certain grouping is present in a molecule by measuring the intensity of the characteristic vibrational band.

In molecules where there is considerable resonance or delocalization, the positions of the vibrational lines can be significantly changed, and this often makes the interpretation of the spectrum much more difficult.

Molecule	Bond	E (cm^{-1})
	Triple bonds	
HCCH	C≡C	1960
CO	C≡O	2150
RCN	C≡N	2250
	Double bonds	
H_2CCH_2	C=C	1620
R_2CO	C=O	1720
CH_3CHNOH	C=N	1650
CH_3ONO	N=O	1640
	Single bonds	
H_3CCH_3	C—C	990
CH_3OH	C—O	1030
CH_3NH_2	C—N	1030
NH_2OH	N—O	1000

4.6 THE ROTATION OF POLYATOMIC MOLECULES

The rotation of polyatomic molecules is a very complex phenomenon. The number of degrees of freedom for most molecules is three, and this involves three independent components of the moment of inertia. The understanding of rotational spectra is based on the relations which exist between the various moments of inertia. In the case of diatomic molecules the interpretation of the rotational spectrum was straightforward because there is only one moment of inertia.

If we assume that the change in the moment of inertia in a vibrating molecule and the interaction of rotation and vibration can be neglected, the rotational energy can be separated from the vibrational energy. Indeed, if incident radiation in the microwave region is used, the rotational levels alone are excited and a pure rotation spectrum is obtained.

The rotational energy of a molecule is given by

$$E = \frac{1}{2}\left(\frac{M_x^2}{I_x} + \frac{M_y^2}{I_y} + \frac{M_z^2}{I_z}\right) \tag{35}$$

where M_x, M_y and M_z are the components of angular momentum resolved along the x, y and z axes and I_x, I_y and I_z are the resolved components of the moment of inertia.

The solutions of (35) are classified according to the relations which exist between the moments of inertia. If $I_x = I_y$ and $I_z = 0$, the molecule is linear; the second case arises when all three moments of inertia are the same: $I_x = I_y = I_z$. This occurs when the molecule in question has two or more two-fold rotation axes of symmetry, and the best known examples of this sort, termed "spherical tops", are molecules with tetrahedral symmetry like CH_4, CCl_4, etc., and also molecules which have octahedral symmetry like SF_6. When $I_x = I_y$ and $I_z \neq 0$, the rotor is known as a "symmetric top", and molecules of this type possess one three-fold axis of symmetry. Examples of symmetric tops are the molecules CH_3Cl, NH_3 and PH_3. The final type of rotor is the "asymmetric top", in which the moments of inertia are all non-zero and all different, and many molecules fall into this class, e.g. CWXYZ.

Molecules may have some of their moments of inertia equal, either by symmetry or else accidently, and the classification into rotor type makes no differentiation between them. The various types of rotor have different degeneracies in their energy levels. The following table lists the four rotor types together with the equation for their various energy levels and the degeneracies of these divers levels.

Rotor type	Energy level	Degeneracy
Linear	$bJ(J+1)$	$2J+1$
Spherical	$bJ(J+1)$	$(2J+1)^2$
Symmetric	$bJ(J+1)+(a-b)K^2$	$(2J+1)$ if $K=0$ and
Asymmetric	no closed form	$2(2J+1)$ if $K \neq 0$

Where the total angular momentum has the value $\{J(J+1)\}^{\frac{1}{2}}(h/2\pi)$ and K is the component of the angular momentum about the z axis and has the eigenvalue $K(h/2\pi)$. The constant b, the inertial constant, is given by $h^2/8\pi^2 I$. For the symmetric top we have two inertial constants corresponding to the two distinct moments of inertia.

Measurement of rotational quanta enable the moments of inertia of the molecule to be calculated and hence also the molecular geometry. The development of microwave spectroscopy since the last war has enabled this to be done with great accuracy.

4.7 ELECTRONIC SPECTRA

There is little to add to the discussion of the electronic spectrum of the diatomic molecule except to say that the spectrum is now much more complicated. Not only are there progressions of vibrations of one type in the electronic band, but there may be several progressions, corresponding to different vibrations superimposed upon each other. Coupled with this there are the rotational lines which accompany each vibrational band. The analysis of the electronic spectrum of a polyatomic molecule is one of great complexity since there are often thousands of lines to classify in the interpretation of perhaps only a single electronic transition.

Like the infrared spectrum, the ultraviolet spectrum of a complicated molecule often shows electronic bands in places which are characteristic of individual groups. This often enables ultraviolet spectroscopy to be used in the same "fingerprint" way as infrared.

In conclusion, it may be noted that there are often significant interactions between vibrations and electronic states. In a linear molecule where the electronic state is degenerate, suitable vibrations will interact to break the degeneracy. This is known as the Renner effect and is betrayed by the presence of line-doubling in a band. Another important effect in non-linear molecules occurs when the electronic state is degenerate. The molecule will either distort itself through vibrations to remove the degeneracy by changing the molecular geometry or else will tend to become linear. This effect (of considerable importance) is known as the Jahn–Teller effect and is of great importance in the spectra of inorganic complexes.

5
Conjugated and Aromatic Molecules

5.1 INTRODUCTION

Molecules in which there are a large number of possible valence bond structures for the ground state have electronic structures which cannot be interpreted in terms of the predominance of any one of them. These are electronic systems in which there is considerable resonance and a large measure of stabilization. In terms of the molecular orbital theory, these molecules can be understood to involve extensive delocalization of some of the electrons. This is of greatest significance in that group of organic molecules which contain an extended chain of atoms, all with p electrons in orbitals perpendicular to the plane of the molecule. These electrons form π orbitals.

Consider the case of *trans*-butadiene, shown below in figure 19.

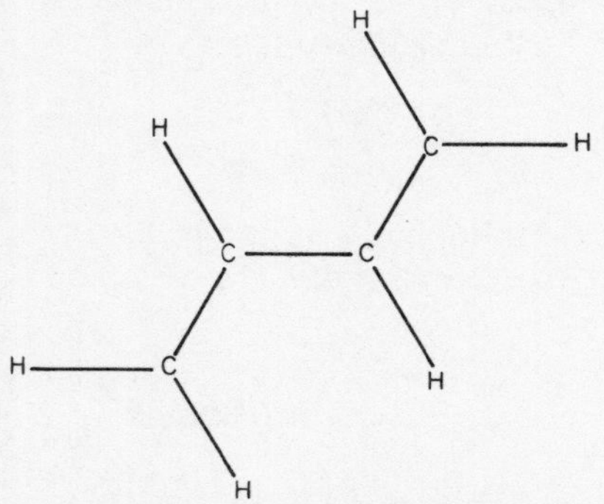

FIGURE 19. *trans*-Butadiene.

The carbon atoms are all in a valence state in which there are $3sp^2$ hybrids associated with each atom, each containing one electron. These orbitals combine either among themselves or with $1s$ orbitals of hydrogen atoms to form a flat molecule in which all of the bond angles are 120°. As with ethylene, there is one electron remaining in a p orbital centred on each carbon atom. These p electrons overlap feebly (S_{ij} between nearest neighbours has an average value of about 0·25 for carbon $2p$ orbitals with an internuclear distance of 1·40 Å), forming π orbitals which extend over the whole molecular framework. This delocalization of the p electrons leads to considerable stabilization of the molecule. The resulting stabilization energy, with respect to a molecule with fixed double bonds, is known as resonance energy in the context of valence bond theory and delocalization energy in molecular orbital theory.

This approach also assumes that the p electrons can be considered independent of the σ bonded framework. This model was first proposed by E. Hückel in 1931, and is known as the Hückel approximation.

In what follows we shall consider both the valence bond and the molecular orbital theories in their description of the behaviour of conjugated and aromatic molecules.

5.2   THE MOLECULAR ORBITAL THEORY

The molecular orbital ψ_J can be written as a linear sum of the participating p orbitals, ϕ_r, centred upon atom r, s etc. throughout the conjugated chain. The molecular orbital ψ_J is given by

$$\psi_J = \sum_r c_{r,J}\phi_r \tag{1}$$

Taking this as a linear variation function, we are required to find the coefficients $c_{r,J}$ which minimize the integral

$$\frac{\int \psi_J^* h\psi_J \, d\tau}{\int \psi_J^* \psi_J \, d\tau} \tag{2}$$

h is the effective Hamiltonian, and in the context of the Hückel theory, the values of the matrix elements of h are

$$h_{ab} = \int \phi_a^* h\phi_b \, d\tau = \beta \quad \text{(where a and b are nearest}$$
$$\text{neighbours, zero otherwise)}$$
$$h_{aa} = \int \phi_a^* h\phi_a \, d\tau = \alpha \tag{3}$$

α and β are known as the Coulomb and resonance integrals respectively and are both negative quantities. The Coulomb integral represents the interaction of a p electron with its own screened nucleus and the other screened nuclei. The integral β represents the interaction between an electron partially located in ϕ_a and partially in ϕ_b with the σ-bonded framework. These quantities are evaluated using experimental data.

In this approximation the interaction of the π electrons among themselves is not explicitly included. α and β are considered independent of their environment and of the π electron distribution. This is of course only roughly valid, and more sophisticated theories must allow this variation both with environment and with the π-electron density.

Insertion of (1) in (2) leads to the following set of simultaneous equations for stationary points in the energy.

$$\sum_s (h_{rs} - S_{rs}E)c_s = 0 \qquad (r = 1, 2, \ldots n) \tag{4}$$

Non-trivial solutions of (4) occur only when

$$|h_{rs} - S_{rs}E| = 0 \tag{5}$$

This leads to n eigenvalues (or orbital energies) and substitution of the E_J one at a time into (4) leads to the atomic orbital coefficients c_{sJ}, normalization being achieved through the relation

$$\sum_{\substack{rs \\ r < s}} (c_{rJ}^2 + 2c_{rJ}c_{sJ}S_{rs}) = 1 \tag{6}$$

In practice, it is often assumed that the atomic orbitals form an orthogonal set, i.e. $S_{ab} = 0$ (unless a = b). Whilst this is true only to a very rough degree, orbitals can be constructed out of the $2p$ orbitals which are rigorously orthogonal. If the orbitals are orthogonal, then (4), (5) and (6) become

$$\sum_s (h_{rs} - S_{rs}E)c_s = 0 \qquad (S_{rs} = 1 \text{ if } r = s, 0 \text{ otherwise}) \tag{7}$$

$$|h_{rs} - S_{rs}E| = 0 \tag{8}$$

$$\sum_r c_{rJ}^2 = 1 \tag{9}$$

Using this approach let us calculate the π molecular orbitals and energies for butadiene. Label the $2p$ orbitals ϕ_a, ϕ_b, ϕ_c and ϕ_d for orbitals centred upon the carbon atoms going from right to left in figure 19.

The Hamiltonian for this system using (3) is

$$\begin{Bmatrix} \alpha & \beta & 0 & 0 \\ \beta & \alpha & \beta & 0 \\ 0 & \beta & \alpha & \beta \\ 0 & 0 & \beta & \alpha \end{Bmatrix} \tag{10}$$

The corresponding secular determinant is

$$\begin{vmatrix} \alpha-E & \beta & 0 & 0 \\ \beta & \alpha-E & \beta & 0 \\ 0 & \beta & \alpha-E & \beta \\ 0 & 0 & \beta & \alpha-E \end{vmatrix} = 0 \tag{11}$$

It is a computational convenience to divide through by β and write $(\alpha-E)/\beta = -x$, so that x is the energy in units of β with respect to the reference zero α (which represents the energy of a $2p$ electron on a carbon atom in its sp^2 valence state). (11) then becomes

$$\begin{vmatrix} -x & 1 & 0 & 0 \\ 1 & -x & 1 & 0 \\ 0 & 1 & -x & 0 \\ 0 & 0 & 1 & -x \end{vmatrix} = 0 \tag{12}$$

Expansion of the determinant and solution of the resultant polynomial equation in x leads to the following eigenvalues

$$x = \pm 0{\cdot}618, \qquad x = \pm 1{\cdot}618 \tag{13}$$

These correspond to orbital energies

$$\begin{aligned} E_1 &= \alpha+1{\cdot}618\beta, & E_2 &= \alpha+0{\cdot}618\beta \\ E_3 &= \alpha-0{\cdot}618\beta, & E_4 &= \alpha-1{\cdot}618\beta \end{aligned} \tag{14}$$

The molecular orbitals which correspond to these eigenvalues are found to be

$$\begin{aligned} \psi_1 &= 0{\cdot}3717(\phi_1+\phi_4)+0{\cdot}6015(\phi_2+\phi_3) \\ \psi_2 &= 0{\cdot}6015(\phi_1-\phi_4)+0{\cdot}3717(\phi_2-\phi_3) \\ \psi_3 &= 0{\cdot}6015(\phi_1+\phi_4)-0{\cdot}3717(\phi_2+\phi_3) \\ \psi_4 &= 0{\cdot}3717(\phi_1-\phi_4)-0{\cdot}6015(\phi_2-\phi_3) \end{aligned} \tag{15}$$

The total energy of the molecule is given by

$$E = 2 \sum_J E_J \tag{16}$$

where the electrons are fed two at a time into the available orbitals until all of the electrons are accommodated. This implies that in the usual case where there are n atomic orbitals, each supplying one electron, the lowest $n/2$ molecular orbitals will be occupied.

It will be noted from (14) and (15) that the orbital energies are placed symmetrically about α and also that there is a relation between the orbital coefficients of orbitals having energies $\alpha + x_r\beta$ and $\alpha - x_r\beta$. This behaviour is characteristic of an important class of aromatic or conjugated molecules called "alternants" by Coulson and Rushbrooke. Consider the molecule

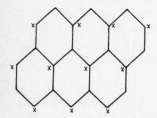

FIGURE 20. An alternant hydrocarbon.

then commencing at any atom and starring every alternate atom, we find that irrespective of where we start, the atoms always fall into the same two groups, every atom being surrounded only by atoms of the other group. This is the defining property of alternants. Alternants which have an even number of carbon atoms have some interesting properties:

1. The eigenvalues (orbital energies) of the secular determinant are symmetrically placed about α, i.e. if $\alpha + x_i\beta$ is an eigenvalue then $\alpha - x_i\beta$ is also. Every bonding orbital has an antibonding partner.
2. If the coefficients of a bonding orbital are $a_1, a_2, \ldots,$ and $b_1, b_2, \ldots, a_i$ and b_i being coefficients of atomic orbitals which belong to the starred and unstarred sets respectively, then the atomic orbital coefficients of the related anti-bonding orbital are $a_1, a_2, \ldots, -b_1, -b_2, \ldots.$

The physical significance of the atomic orbital coefficients in (15) is readily determined. The probability density of finding an electron at any point r in space due to an electron in molecular orbital ψ_J is given by $\psi_J^*(r)\psi_J(r)$. The total probability or electron density due to all electrons in all occupied orbitals is

$$P = 2 \sum_J \psi_J^*(r)\psi_J(r) \tag{17}$$

Substituting (1) in (17) gives

$$P = \sum_r P_{rr}\phi_r^*\phi_r + \sum_{\substack{r,s \\ r \neq s}} P_{rs}\phi_r^*\phi_s \tag{18}$$

where

$$P_{rr} = 2 \sum_J c_{rJ}c_{rJ} \quad \text{and} \quad P_{rs} = 2 \sum_J c_{rJ}c_{sJ} \tag{19}$$

P_{rr} is known as the π-charge density or π charge on atom r, and P_{rs} is the π bond order of the bond r–s and is proportional to the amount of charge due to the π electrons to be found between atoms r and s.

Introducing these charges and bond orders into our discussion of alternants leads to two further properties of alternants containing an even number of carbon atoms.

3. If each atom contributes one π-electron to the molecule then the charge distribution is uniform, each P_{ii} having the value unity.
4. Bond orders between atoms of the same set are zero.

In butadiene, which we examined above, we find

$$P_{11} = P_{22} = P_{33} = P_{44} = 1 : P_{12} = P_{34} = 0.894$$

$$P_{23} = 0.447, \qquad P_{14} = -0.447 \quad \text{and} \quad P_{13} = P_{24} = 0$$

assuming that the lowest two orbitals are doubly occupied. We note that the three bonds in butadiene all possess some double bond character, this being greatest in the case of the two outer bonds. It is this delocalization which leads to the stability of the molecule.

5.3 BOND LENGTHS AND BOND ORDERS

We noted above that the bond order is a measure of the π-electron charge which is localized in the bond area between the atoms concerned.

This is easily seen by integrating (18) over all space. This gives

$$\int P \, dv = \sum_r \int P_{rr} \phi_r^* \phi_r \, dv + \sum_{r,s} \int P_{rs} \phi_r^* \phi_s \, dv$$

$$= \sum_r P_{rr} \int \phi_r^* \phi_r \, dv + \sum_{r,s} P_{rs} \int \phi_r^* \phi_s \, dv \qquad (20)$$

$$= \sum_r P_{rr} + \sum_{\substack{rs \\ r \neq s}} P_{rs} S_{rs} = N$$

where N is the total number of π electrons.

Since the overlap integral is constant to a first approximation for small variations in bond length, it can be seen that the bond order is a very real measure of the charge in the bond region. Bond order is defined not only for nearest neighbours but for more distant neighbours, for example, in benzene the bond order between atoms *para* to each other is $-\frac{1}{3}$.

That bond lengths vary with the degree of double bond character in the bond is obvious on a little reflection. Benzene, which has a bond order of 0·667, has a bond length of 1·39 Å, whilst ethane, which has no double bond character ($P_{ij} = 0$), has a bond length of 1·54 Å and ethylene, which has a fixed double bond or 100% double bond character ($P_{ij} = 1·000$), has a bond length of 1·32 Å. The bond length then varies with the π-bond order. Coulson found a relationship of the type

$$x = s - \frac{s-d}{1 - kP/(P+1)} \qquad (21)$$

where x is the bond length of the bond whose π-bond order is P; s is the length of a "natural" single bond and d the length of a "natural" double bond. k is a constant whose value must be determined from empirical data. In fact, it is not a good idea to determine the "natural" single bond length from (say) ethane, since this involves the carbon atom in a different valence state; it is best to regard this as another empirical parameter. Ethylene and benzene supply two points on a curve whilst graphite, where the bond orders can be easily determined, can be used as a third point. From this data it is found that s has the value 1·515 Å (slightly shorter than the ethane value) and k has the value 1·05 Å. Bond orders predicted using this relationship are found to be in fair agreement. For naphthalene and anthracene they are shown in the first table on page 120.

From this table it can be seen that there is no deviation greater than 0·03 Å. This appears to be about the poorest degree of accuracy with which bond orders can be predicted using this method, although

Naphthalene		Bond length (Å)			
Bond	1–2	2–3	3–8	1–10	
Predicted	1·34	1·41	1·42	1·40	
Observed	1·37	1·43	1·39	1·40	

Anthracene		Bond length (Å)				
Bond	1–2	2–3	3–4	3–12	1–14	
Predicted	1·38	1·41	1·40	1·42	1·40	
Observed	1·37	1·42	1·40	1·44	1·41	

in many cases much better agreement is obtained. It appears that the larger the molecule the better is the agreement between predicted bond lengths and empirical values.

5.4 RESONANCE ENERGY

As was noted above, conjugated molecules possess a lower energy than they would if the structure consisted of isolated single and double bonds. This extra stabilization energy is known as delocalization or resonance energy. In the case of benzene it is found that the delocalization energy is of the order of 64 kcal mole^{-1} with respect to the corresponding molecule in which the double bonds are localized.

Carrying through a molecular orbital calculation for benzene similar to that discussed above for butadiene, we obtain the following results:

Orbital energy	Orbital
$\alpha+2\beta$	$\psi_1 = \dfrac{1}{\sqrt{6}}(\phi_1+\phi_2+\phi_3+\phi_4+\phi_5+\phi_6)$
$\alpha+\beta$	$\psi_2 = \dfrac{1}{2}(\phi_2+\phi_3-\phi_5-\phi_6)$
$\alpha+\beta$	$\psi_3 = \dfrac{1}{\sqrt{12}}(2\phi_1+\phi_2-\phi_3-2\phi_4-\phi_5+\phi_6)$
$\alpha-\beta$	$\psi_4 = \dfrac{1}{\sqrt{12}}(2\phi_1-\phi_2-\phi_3+2\phi_4-\phi_5-\phi_6)$

Orbital energy	Orbital
$\alpha - \beta$	$\psi_5 = \frac{1}{2}(\phi_2 - \phi_3 + \phi_5 - \phi_6)$
$\alpha - 2\beta$	$\psi_6 = \frac{1}{\sqrt{6}}(\phi_1 - \phi_2 + \phi_3 - \phi_4 + \phi_5 - \phi_6)$

The total energy of the molecule is $6\alpha + 8\beta$. If the molecule contained three isolated single bonds and three isolated double bonds, the total energy would be $6\alpha + 6\beta$. The delocalization energy is 2β. More elaborate treatment of the empirical data yields a value of 36 kcal mole^{-1} for the resonance energy. From this data we have a method for the determination of the resonance integral β.

In the table below are listed the resonance energies of the lower members of the polyacene series together with their calculated delocalization energies.

Hydrocarbon	Delocalization energy	Resonance energy (kcal mole^{-1})
Benzene	$-2 \cdot 000$	$36 \cdot 0$
Naphthalene	$-3 \cdot 683$	$61 \cdot 0$
Anthracene	$-5 \cdot 314$	$83 \cdot 5$
Naphthacene	$-6 \cdot 932$	$110 \cdot 0$

These data give for β the value about 16·5 kcal mole^{-1}, although there is variation from 18 for benzene to 15·9 for naphthacene. For the polyene series a value of about 6 kcal mole^{-1} is obtained.

In deriving the value for the delocalization energy for a molecule we have assumed a hypothetical reference model in which all bonds are equal, irrespective of whether they are pure single or double bonds. This reference model is obtained from the "real" model with isolated bonds by considerable deformation of the bond lengths, requiring a substantial amount of energy involved with compression and extension of the bonds. Elaborate calculations have been made to estimate the true value of the delocalization energy. The values of β obtained in this way vary between 31·5 to 55·5 kcal mole^{-1}. The delocalization energy which is calculated in the way discussed above is known as a "vertical resonance energy".

Calculations of this type are extremely difficult, since compression energies are not easy to estimate, and correction to the estimated vertical resonance energy to refer to the molecule with varying bond lengths is an unsure procedure.

5.5 *HETEROCYCLICS*

The theory discussed above can be extended to the molecules in which conjugation still occurs but in which C—H groups are replaced by —N— atoms. The simplest example is the replacement of one C—H in benzene to give pyridine, as shown in figure 21.

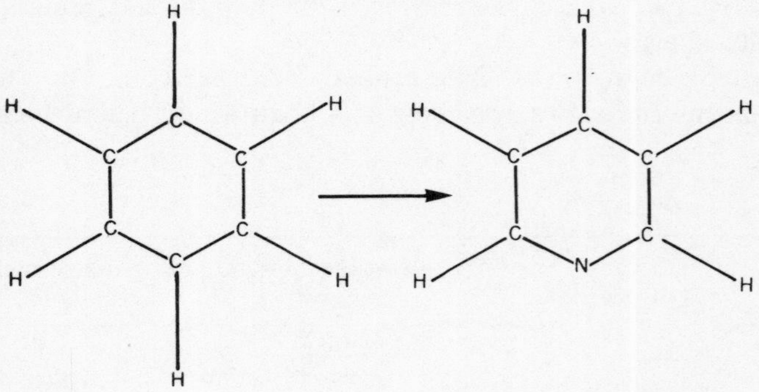

FIGURE 21.

In making such a substitution it is necessary to revise the values of α and β. In the case of the Coulomb integral, α is replaced by $\alpha + \beta\delta$ in order to describe the difference in electronegativity between the substituent atom and the displaced carbon atom. The value of δ is chosen to fit experimental data. In the case of nitrogen atoms it used to be fashionable to use values of $2 \cdot 0\beta$ for $\beta\delta$, but more recent work puts this value much lower, probably between $0 \cdot 30$ and $0 \cdot 50\beta$.

The value of the resonance integral β for C—N bonds varies so slightly from the C—C value that it is conventional to assign the same value to it.

When we come to oxygen heterocyclics, values of δ used vary considerably. For molecules containing the C=O group, where the oxygen atom contributes one electron to the conjugated system, values of δ vary between about $0 \cdot 6$ and $1 \cdot 0$, which is in fair agreement with electronegativity considerations as we go from carbon to nitrogen to oxygen.

5.6 DIPOLE MOMENTS

When a heteroatom is introduced into an aromatic ring or a conjugated chain, the charge distribution ceases to be uniform. At first sight it would appear that calculation of dipole moments would give an excellent way of estimating the value of δ.

The dipole moment of a molecule is given by

$$\boldsymbol{\mu} = e \int \psi^* \mathbf{r} \psi \, dv = e \left\{ \mathbf{i} \int \psi^* x \psi \, dv + \mathbf{j} \int \psi^* y \psi \, dv + \mathbf{k} \int \psi^* z \psi \, dv \right\} \quad (22)$$

where e is the charge on the electron. As noted before the dipole moment is a vector quantity and its magnitude is given by

$$\boldsymbol{\mu} = (\boldsymbol{\mu}^2)^{\frac{1}{2}} = (\mu_x^2 + \mu_y^2 + \mu_z^2)^{\frac{1}{2}} \quad (23)$$

The dipole moments which arise from the σ electrons in conjugated molecules may be estimated by vector addition of the individual bond moments, since these are relatively constant constitutive properties of the σ bonds. The total dipole moment is

$$\boldsymbol{\mu} = \boldsymbol{\mu}^\sigma + \boldsymbol{\mu}^\pi \quad (24)$$

The π dipole moment can be evaluated in the following way. Since the dipole operator is a one-electron operator, the total π moment is given by

$$\boldsymbol{\mu} = e2 \sum_J \int \psi_J^* \mathbf{r} \psi_J \, dv = e2 \sum_J \sum_{rs} c_r^* c_s \int \phi_r^* \mathbf{r} \phi_s \, dv \quad (25)$$

where

$$r = x\mathbf{i} + y\mathbf{j} + z\mathbf{k}$$

The summed products of coefficients are the charges and bond orders P_{ii} and P_{ij}. Multiplying and dividing each term by the appropriate overlap integral $\int \phi_r^* \phi_s \, dv$, we obtain

$$\mu = e \left(\sum_r P_{rr} \frac{\int \phi_r^* \mathbf{r} \phi_s \, dv}{\int \phi_r^* \phi_s \, dv} \, dv \int \phi_r^* \phi_r \, dv + \sum_{r,s} P_{rs} \frac{\int \phi_r^* \mathbf{r} \phi_s \, dv}{\int \phi_r^* \phi_s \, dv} \int \phi_r^* \phi_s \, dv \right) \quad (26)$$

The first half of each term is the component of the centroid of the atomic orbital r or of the bond r–s. Equation (26) then becomes

$$\mu_x = e \left(\sum_r P_{rr} X_r + \sum_{r,s} P_{rs} X_{rs} S_{rs} \right) \quad (27)$$

with similar expressions for the y and z components.

The ionic framework will also have a dipole moment given by

$$\mu_x^I = \sum_r Z_r X_r \tag{28}$$

where Z_r is equal to the number of π-electrons donated to the delocalized system by the atom r. Similar expressions hold for the y and z components. The contribution of the π electrons to the total dipole moment will be given by the difference between (27) and (28). Writing the net π charges as

$$q_r = (P_{rr} - Z_r) \tag{29}$$

and

$$q_{rs} = 2P_{rs}S_{rs} \tag{30}$$

and setting the electronic charge to unity we obtain for the x component

$$\mu_x^\pi = \sum_r q_r X_r + \sum_{\substack{rs \\ r>s}} q_{rs} X_{rs} \tag{31}$$

The total π moment is given by

$$\mu^\pi = (\mu_x^{\pi^2} + \mu_y^{\pi^2} + \mu_z^{\pi^2})^{\frac{1}{2}} \tag{32}$$

The centre of mass of the molecule is usually regarded as the origin of the coordinates in which the atom and bond centroids are expressed.

When the overlap integrals are assumed to be zero (31) becomes

$$\mu_x^\pi = \sum_r q_r X_r \tag{33}$$

For the nitrogen heterocyclics, pyridine, pyridazine, pyrimidine and pyrazine

Pyridine   Puridazine   Pyrimidine   Pyrazine

we have the following values for the π dipole moments

	Calculated	Observed
Pyridine	2·30 D	2·2 D
Pyridizine	3·69	3·9
Pyrimidine	2·26	2·4
Pyrazine	0·00	0·00

The units are 10^{-18} e.s.u., which are known as Debyes (D). These values are obtained using a value of $0·58$ for δ. Similar calculations on oxygen heterosystems (where the oxygen atom contributes one π electron) yield a value of $1·0$ for δ.

If the coordinates of the centroids are expressed in angstroms (cm^{-8}) and the electronic charge is regarded as unity, (31) becomes

$$\mu_x^\pi = 4.77\left(\sum_r q_r X_r + \sum_{rs} q_{rs} X_{rs}\right) D \tag{34}$$

Values of δ estimated from dipole moment data are not very reliable because they involve an estimation of the dipole moment for the lone pairs of electrons on the heteroatoms which are part of the σ framework. These can be partially taken into account, by determining the dipole moments of a series of molecules with respect to one member (say pyridine in our case).

5.7 IONIZATION POTENTIALS

The total energy of a molecule in which each molecular orbital is doubly occupied until all of the available electrons has been accommodated is

$$E = 2\sum_J E_J \quad \text{(over all occupied orbitals)} \tag{35}$$

When one electron is removed, the topmost orbital, say ψ_K, will be singly occupied and the total π-electron energy for the ion is

$$E^+ = 2\sum_J E_J + E_K \qquad (J \neq K) \tag{36}$$

The difference in energy between the molecule and the ion, is that energy which is required to ionize the molecule and is called the ionization potential, is given by

$$I = E^+ - E = -E_K \tag{37}$$

This statement that the ionization potential is equal to minus the energy of the top filled orbital is known as Koopman's theorem.

Values for the ionization potential enable an estimate of both α and β to be made. In the case of graphite the topmost orbital has an energy α. The ionization potential of graphite is $4.3\,eV$. The value of α is $-4.3\,eV$. Having determined α from the datum for graphite, the value of β can be obtained from an examination of the ionization potentials and orbital energies of a series of molecules. Using the polyacene series we have

	I	E_K
Benzene	$9.57\,eV$	$\alpha + 1.000\beta$
Naphthalene	8.68	$\alpha + 0.618\beta$
Anthracene	8.20	$\alpha + 0.414\beta$
Naphthacene	7.71	$\alpha + 0.295\beta$

Plotting I against $-E$, a value of about $-2\cdot4$ eV (or -55 kcal mole^{-1}) is obtained for β. There is however a wide scatter in the plot. These values are not in very good agreement with the value for β obtained from resonance energy measurements.

5.8 FREE VALENCE AND REACTIVITY

The effective number of electrons around an atom can be regarded as a measure of its reactivity towards positively charged ions. The net charges q_r do not represent the total amount of π-electron charge in excess about a given atom. Indeed, for alternant hydrocarbons the q_r are all zero, and we would thus expect all hydrocarbons to be unreactive. This is not the case.

A further contribution to the π-electron density in the neighbourhood of an atom comes from the amount of charge in the overlap region between an atom and its neighbours. Since, to first order, the overlap integral is constant between nearest neighbours, the bond order is a measure of this charge. If this charge reaches a certain critical value the atom will have its valence requirements satisfied. The sum of the bond orders around a given atom will therefore be a measure of the extent to which the valence requirements are satisfied, and also an inverse measure of the ability of the atomic site to react with a ligand. Coulson defined a number N_i known as the bond number, which is the sum of the bond orders (σ as well as π) about the atom i.

Consider butadiene

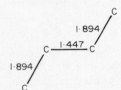

FIGURE 22. Bond orders for butadiene.

(Figures refer to total bond orders $P_{rs} = P_{rs}^{\sigma} + P_{rs}^{\pi}$.)

The bond order sum for the outer atoms is made up of contributions from the three σ bonds which the atom forms (with the adjacent carbon atom and with the two hydrogen atoms to which it is attached) and from the π bond between the carbon atoms which has a bond order of $0\cdot894$. The quantity $N_1 = N_4$ has the value

$$3\cdot000 + 0\cdot894 = 3\cdot894$$

and for the internal carbon atoms

$$N_2 = N_3 = 3{\cdot}000 + 0{\cdot}894 + 0{\cdot}447 = 4{\cdot}341$$

The inner carbon atoms are much more involved than the outer carbon atoms in bonding, and it is therefore to be expected that the terminal atoms will be more reactive than the inner atoms. In practice this is found to be the case.

Coulson has shown that the maximum possible value for N_i (excluding triple bonds) is $4{\cdot}732 = N_{max}$. The reactivity of an atomic site can be represented by

$$F_r = N_{max} - N_r \qquad (38)$$

This quantity F_r is known as the free valence of the atom r. For butadiene it has the value $0{\cdot}838$ for the terminal carbons and $0{\cdot}391$ for the inner carbon atoms. This quantity is the wave-mechanical analogue of Thiele's partial valencies.

Free valencies are usually represented as numbers placed at arrow heads directed away from the atom concerned. In figure 23 free valencies are shown for a few representative conjugated and aromatic molecules.

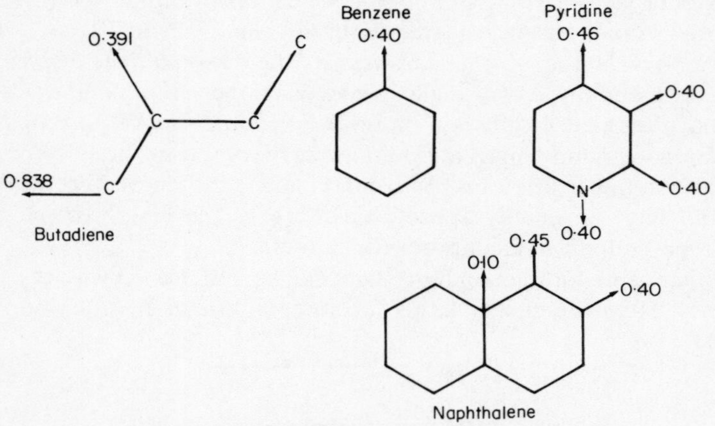

FIGURE 23. Free valences for selected molecules.

5.9 ELECTRONIC SPECTRA

An electronic transition, giving rise to a band in the ultraviolet region of the spectrum, can be approximately described by promoting electrons from orbitals which are occupied in the ground state to orbitals which are vacant in the ground state. The energy of the orbital from which the electron jumps (E_J) is, say, $E\alpha + x_J\beta$, and the energy of the orbital to which it goes has a value $E_K = \alpha + x_K\beta$. The energy of the transition is given by

$$E_{J \to K} = E_K - E_J = \alpha + x_K\beta - (\alpha + x_J\beta) = (x_K - x_J)\beta \qquad (39)$$

Correlation of empirical spectroscopic data with molecular orbital calculations provides another method for the determination of the resonance integral β.

In the table which follows, differences in orbital energy between the bottom unfilled orbital and the highest filled orbital are given together with the energy of the lowest band in the electronic spectrum, for the first three members of the polyene series.

	Energy (cm^{-1})	$E_K - E_J$ (β)
Ethylene	61,500	$-2 \cdot 000$
Butadiene	46,080	$-1 \cdot 236$
Hexatriene	39,750	$-0 \cdot 890$

This data gives a value of about $-4 \cdot 7$ eV for β, but as we go up the series the ratio of obs./calc. increases, in this case from $30{,}750 \text{ cm}^{-1}/\beta$ for ethylene to $44{,}500 \text{ cm}^{-1}/\beta$ for hexatriene. The reason for the discrepancy is that the assumption of a single β depends on the validity of the assumption that the bond lengths are more or less equal. In the polyene series this is not so; bond-length alternation persists even in the polyene with an infinite number of atoms. As a consequence the lowest spectral band does not have an energy approaching zero as the length of the chain becomes infinite; it in fact approaches a limit.

Another cause for the non-agreement can be understood with the aid of figure 24. Here the energy levels for some alternant hydrocarbons are shown:

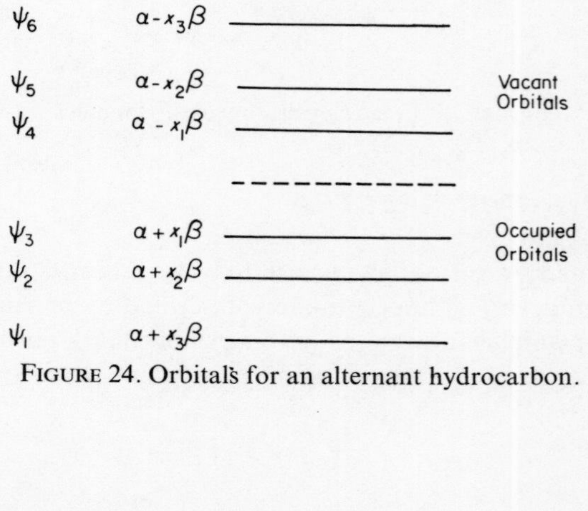

$\psi_6 \qquad \alpha - x_3\beta$ ———————

$\psi_5 \qquad \alpha - x_2\beta$ ———————   Vacant Orbitals
$\psi_4 \qquad \alpha - x_1\beta$ ———————

- - - - - - - - - -

$\psi_3 \qquad \alpha + x_1\beta$ ———————   Occupied Orbitals
$\psi_2 \qquad \alpha + x_2\beta$ ———————

$\psi_1 \qquad \alpha + x_3\beta$ ———————

FIGURE 24. Orbitals for an alternant hydrocarbon.

From an examination of the figure it can be seen that the transition from ψ_3 to ψ_5 has the same energy as the transition from ψ_2 to ψ_4. This means that the transition is degenerate and the true state of affairs is obtained by taking linear combinations $(\psi_{3\to 5} + \psi_{2\to 4})$ and $(\psi_{3\to 5} - \psi_{2\to 4})$. These two states will have different energies because in general there will be a matrix element $h_{3\to 5,2\to 4}$ which is non-vanishing. The energies of the two transitions will be given by $E + h_{3\to 5,2\to 4}$ and $E - h_{3\to 5,2\to 4}$. This phenomenon is known as configuration interaction and is not confined to the interaction of degenerate configurations. Any two excited configurations of the same symmetry will, in general, interact.

In benzene, if we consider the transitions between the bottom two filled and the top two unfilled orbitals (both of which are degenerate pairs), there arise four excited configurations which have the same energy. However they fall into two pairs of differing symmetry, and the resultant states arise from interaction in pairs as shown in figure 25. (A and B represent one degenerate pair of orbitals with energy $\alpha + \beta$ and A' and B' their alternate counterparts with energy $\alpha - \beta$.)

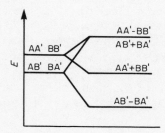

FIGURE 25. Configuration interaction among excited configurations of benzene.

However even when this is included, it is only in very few cases that the theory is able to predict accurately the spectra of a series of related molecules.

5.10 ASSESSMENT OF M.O. THEORY

From what has been said above, it can be seen that the molecular orbital theory can explain in a qualitative way the properties of this class of molecules. Indeed the basic concept of the Hückel model as a sea of π-electrons moving freely over the tightly-bonded σ framework can be demonstrated convincingly by measurements of diamagnetic susceptibility, first in a direction parallel to the plane of the molecule and then perpendicular to it. The value in a direction perpendicular to the plane of the molecule is much greater than the other. This can be understood in terms of the presence of freely-moving π electrons setting up ring currents around the aromatic framework, giving rise to an enhanced diamagnetic susceptibility.

It is possible to make the molecular orbital theory more quantitative by admitting interactions between π electrons and allowing a correct dependence of the Hamiltonian upon the electron distribution in atoms and bonds. This can be done and the main features of the Hückel theory maintained. There is then a problem of self-consistency between the Hamiltonian and the wave function (i.e. charges and bond orders in the direct sense). Predictions of molecular properties using self-consistent field theory lead to much better agreement with empirical data.

5.11   VALENCE BOND THEORY

Historically, the valence bond theory has been of considerable importance in understanding the chemical properties of aromatic and heterocyclic molecules. Even before the emergence of quantum mechanics, the concept of resonance between several "canonical structures" was well-established.

In order to account for the chemical properties of benzene, Kekulé over a century ago postulated that the true structure must lie somewhere between

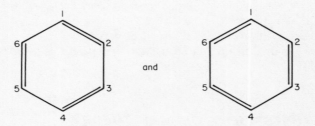

FIGURE 26. Kekulé structures for benzene.

These two structures can be represented by

$$\Psi_1 = \phi_1(1)\alpha(1)\phi_2(2)\beta(2)\phi_3(3)\beta(3)\phi_4(4)\alpha(4)\phi_5(5)\alpha(5)\phi_6(6)\beta(6) \quad (40)$$

which corresponds to perfect pairing (or bonds) between atoms 1 and 2,

3 and 4, and 5 and 6; and the function

$$\Psi_2 = \phi_1(1)\alpha(1)\phi_2(2)\alpha(2)\phi_3(3)\beta(3)\phi_4(4)\beta(4)\phi_5(5)\alpha(5)\phi_6(6)\beta(6) \quad (41)$$

which corresponds to perfect pairing between atoms 2 and 3, 4 and 5, and 6 and 1.

The true wave function is in fact neither one of these possibilities but a linear combination of both of them.

$$\Psi = \frac{1}{\sqrt{2}}(\Psi_1 + c\Psi_2) \quad (42)$$

where c is a constant of modulus unity.

The wave function can be improved by incorporating terms corresponding to further structures. Traditionally, the benzene molecule has been considered as a mixture principally of the two Kekulé structures, together with the three Dewar structures (figure 27) which contain two ordinary nearest-neighbour π bonds and one "long" bond linking the *para* positions.

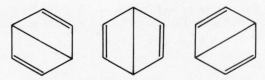

FIGURE 27. Dewar structures for benzene.

The weight of these structures in the total wave function must be determined by carrying through a variational calculation. In order to obtain a better wave function, it is necessary to include in the wave function some of the possible ionic structures. Representative types of ionic structures for benzene are shown in figure 28.

FIGURE 28. Ionic structures for benzene.

In many cases it is found that the only structures which have significant weights in the ground-state wave function are those which correspond to the maximum number of double bonds. Indeed there are many cases where only one structure is possible and in these cases the double bonds will be isolated. In aromatic and other conjugated molecules this is however not so, and the wave function for the groundstate is a linear combination of structures which correspond to different pairings of the electrons. When the wave function can be written in terms of a single structure, the wave function is given by that for the structure, say ψ_A, and the energy is given by

$$E = H_{AA} \tag{43}$$

assuming that the wave function is normalized. H_{AA} can then be represented in terms of a Coulomb integral Q and a set of exchange integrals for nearest neighbours α. These integrals were defined in the discussion of the hydrogen molecule, and are here regarded as empirical quantities. The energy (43) for the structure becomes

$$E = Q + \text{(exchange integrals between bonded orbitals)}$$
$$- \tfrac{1}{2} \text{(exchange integrals between non-bonded orbitals)}$$

There is a ready way of evaluating E, when only nearest neighbour interactions are included. Write down each Kekulé structure without its σ bonds for benzene, e.g. (a):

(a)   (b)

Superimpose this pattern upon itself (b). Each of the double lines is called an "island". A pattern such as (b) consists of a set of closed islands, the number of islands i being three. The matrix element H_{AA} has the value

$$H_{AA} = \frac{1}{2^{N-i}}(Q + b\alpha) \tag{44}$$

where $2N$ is the total number of electrons, Q and α the Coulomb and exchange integrals and b is the number of pairs of nearest neighbours in the islands of the superimposition pattern, minus one-half the number of nearest neighbour orbitals on separate islands. In our example 2^{N-i} has the value unity because N has the value 3 (there being 6π electrons) and $i = 3$, $b = 3 - \tfrac{1}{2} \times 3 = 1\cdot5$. The energy of a Kekulé structure is thus

$Q + 1 \cdot 5\alpha$. For a Dewar structure the energy is Q. These values are the same as found by direct application of the expression for the energy given above.

In order to evaluate the off-diagonal elements of the Hamiltonian H_{AB} between structures, we proceed to construct a superimposition pattern in which one bond pattern is superimposed upon the other. Again a set of islands results, and the element H_{AB} is given by

$$H_{AB} = \frac{1}{2^{N-i}}(Q + b\alpha) \tag{45}$$

where N and i are defined as above and b is the number of nearest neighbour orbitals (whether bonded or not) in the islands minus one half the number of nearest neighbour pairs on separate islands.

Consider the off-diagonal term between (a) two Kekulé structures for benzene, (b) one Kekulé and one Dewar and (c) two Dewar structures. The superimposition patterns for the three cases are shown below:

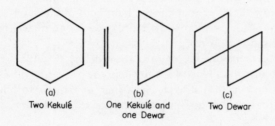

(a)
Two Kekulé

(b)
One Kekulé and
one Dewar

(c)
Two Dewar

Using (45) and the diagrams it is seen that H_{AB} between two Kekulé structures has the value $\frac{1}{4}(Q + 6\alpha)$, between one Kekulé and one Dewar $\frac{1}{2}(Q + 3\alpha)$ and between two Dewar structures $\frac{1}{4}(Q + 8\alpha)$. The modulus of the overlap integral is equal to the coefficient of Q both for diagonal and off-diagonal elements. The sign of S is the same as the sign for Q. Multiplication by the appropriate factor can lead to the set of structures being a normalized set.

It remains to develop a method for enumerating the valence bond structures in some systematic way. In 1932 Rumer showed how to write down a set of structures for a given system. The atoms forming the molecule are written around the circumference of a circle. The numbering of the atoms is immaterial to the method. If there are $2N$ electrons then we start by drawing the structures which contain N bonds. It does not matter whether these bonds link different sorts of atoms, the only restriction being that the line joining any two atoms in the diagram should not cross any line joining another pair of atoms. Having exhausted the N bond case, the process is repeated for the $N-1$ bond case and so on until we arrive at the structure which contains no bonds. When this is reached we have generated the complete set of structures in which only non-polar contributions are included. Any structure involving "crossed" bonds can be written as a linear sum of structures containing only uncrossed bonds.

The most important subset of structures is that which contains those structures having the maximum number of bonds. In the Rumer method of generating the set it is unnecessary to distinguish between nearest and non-nearest neighbour bonds. This difference becomes necessary only when evaluating the elements of the Hamiltonian. The number of structures containing the maximum number of bonds is

$$\frac{N!}{(N/2)!(N/2+1)!}$$

where N is the total number of electrons (which must be even).

This method of generating structures is very important because it ensures that only linearly independent members of the set are generated. The non-crossing rule precludes structures which can be written as linear sums of uncrossed structures.

As an example consider the enumeration of the structures for benzene which contain three bonds. We represent the six $2p$ orbitals as points on the circumference of a circle. The order in which we place them is immaterial, actual neighbours do not have to be neighbours on the circle. In our case we have

To determine the complete set of structures containing three bonds (or three sets of paired orbitals), draw lines between pairs of orbitals such that no two lines cross. We have

These are the only five independent structures for benzene containing three bonds. Any other structure which involves crossed bonds can be written as a linear sum of these five structures.

The set of structures containing two and one bond(s) can be enumerated in the same way.

5.12 RESONANCE ENERGY

When two or more of the contributory structures have either the same or nearly the same energy, the ground state is represented by a linear sum of them all. As we have already said, these structures have no objective reality, and it is only the combination of them which exists. When structures are mixed, the resultant state has an energy which is lower than that of any of the contributing structures. This decrease in energy is known as resonance energy in the valence bond approximation. The variation in the resonance energy for benzene with the inclusion of different structures is shown in the table below.

Structures used	Wave function	Energy	Resonance energy
One Kekulé	ψ_K	$Q + 1.5\alpha$	0.000
Two Kekulé	$\psi_{K1} + \psi_{K2}$	$Q + 2.4\alpha$	0.9α
One Dewar	ψ_D	Q	
Three Dewar	$\psi_{D1} + \psi_{D2} + \psi_{D3}$	$Q + 2.0\alpha$	0.5α
All five	$c_1(\psi_{K1} + \psi_{K2}) + c_2(\psi_{D1} + \psi_{D2} + \psi_{D3})$	$Q + 2.606\alpha$	1.106α

It can be seen from the table that the bulk of the resonance energy in the five-structure wave function comes from the two Kekulé structures. The three Dewar structures cause a gain of only 20% in the resonance energy. It is found that in the wave function the Kekulé structures have a weighting of five to one with respect to the Dewar structures, and thus to a very good approximation benzene can be described in terms of the two Kekulé structures.

The resonance energies of the first three members of the polyacene series (using only fully-bonded structures) are shown below:

	Calculated (α)	Experimental (kcal mole^{-1})
Benzene	1.106	36.0
Naphthalene	2.040	61.0
Anthracene	2.951	83.5

From the data we can estimate the resonance integral α, and it is found to have the value of 30 kcal mole^{-1}. One disturbing feature is that it falls as the number of rings increases; the ratio observed/calculated has the value $32{\cdot}4$, $30{\cdot}0$ and $28{\cdot}3$ for the three members benzene, naphthalene and anthracene respectively. From the table above it can be seen that the resonance energy increases by about $0{\cdot}95$ for every extra ring.

It would seem, that for the polyene series the value of α which accounts for the data would be one of about 15 kcal mole^{-1}.

5.13 FRACTIONAL BOND ORDER

As we have already noted, in aromatic and conjugated systems there are no pure single or pure double bonds, all bonds lying somewhere in between these two extremes. A fractional bond order can be defined within the framework of the valence bond theory from a knowledge of the weight with which each structure occurs in the wave function. Consider any of the bonds in benzene. In either one or the other of the two Kekulé structures this bond appears double and it also does in one of the Dewar structures. Now the fractional weight of each of the Kekulé structures is $0{\cdot}39$, whilst that of each of the Dewar structures is $0{\cdot}07$. The fractional double bond character is therefore $0{\cdot}39 + 0{\cdot}07 = 0{\cdot}46$. In the case of benzene all bonds have the same fractional double bond character.

In the case of butadiene, the outer bonds are found to have $0{\cdot}88$ fractional double bond character whilst the inner bond has a bond character of $0{\cdot}12$.

5.14 FREE VALENCE

Free valence is defined in two different ways in valence bond theory. The simpler of the two alternatives assumes that any electron involved in a Dewar type long bond is free to engage in bond formation with some reacting species. In this definition the free valence is the sum of the weights of structures which have Dewar long bonds at the atom in question. When benzene is considered as the sum of the two Kekulé structures only the free valence of every atom is zero. The wave function which includes the three Dewar structures as well as the two Kekulé structures has a free valence of $0{\cdot}07$ since each Dewar structure has a weighting of 7% and each atom is engaged in only one Dewar-type bond.

The other definition is analogous to that used in molecular orbital theory. The sum of the fractional bond orders is substracted from F_{max} which is given the value 1·732. For benzene, when only Kekulé structures are considered, the value of the free valence F is 0·812, and it has the value 0·740 when the three Dewar structures are included as well as the Kekulé structures.

The values of the free valence for naphthalene are shown below

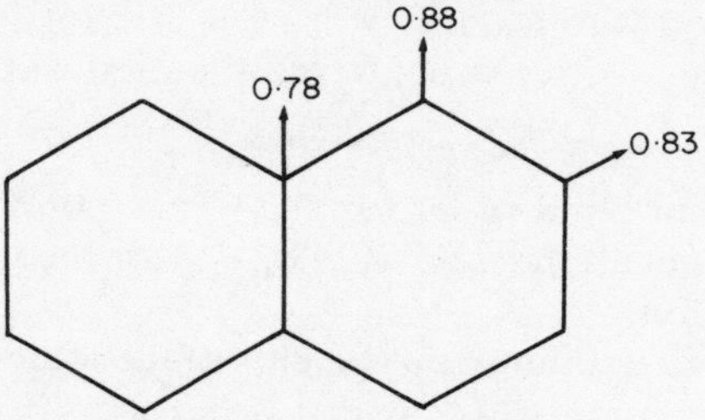

5.15 DIPOLE MOMENTS

In the molecular orbital theory, the presence of a heteroatom involves the change in the Coulomb integral from α to $\alpha + \beta\delta$, where the δ takes into account the difference in electronegativity between the heteroatom and carbon. In the valence bond theory the presence of an heteroatom involves the inclusion of ionic structures in the wave function.

In the case of pyridine, which can be considered as being derived from benzene by replacing one C—H group by N, the total number of π electrons remains unchanged. In addition to the five structures already considered for benzene it is necessary to include some ionic structures, the dominant ones being shown in figure 29.

FIGURE 29. Ionic structures for pyridine.

Inclusion of these structures leads to a flow of electrons onto the nitrogen atom at the expense of the *ortho* and *para* carbon atoms, the *meta* atoms being unaffected. A simple calculation can be carried through using the five covalent structures and these three ionic structures. All that needs to be done is to evaluate a few Q's and α's. This is however a difficult task and it is usually determined in a semi-empirical way.

Let w_K and w_D be the weights of the Kekulé and Dewar structures respectively and the weights of a, b and c be w_a, w_b and w respectively. The charge on the nitrogen atom will be

$$w_a + 2w$$

where the weights are represented as fractions of the total wave function weight. On the ortho atom it will be w and on the para atom w_a. The weighting of the various structures in the total wave function is

$$2w_K + 3w_D + w_a + 2w = 1$$

The fractional bond order for the C—N bond is $w_K + w_D$, $w_K + w_D + w_a + w$ for the next bond, and $w_K + w_D + w$ for the bond most remote from the nitrogen atom.

One way of determining the weights of the ionic structures is to predict a charge distribution which is consistent with the dipole moment data.

It is only on rare occasions that valence bond theory is used to predict dipole moments. It is of much more utility in estimating the weights of the various ionic structures in the total wave function.

5.16   ELECTRONIC SPECTRA

The description of electronic transitions in the valence bond theory is not easy, although it is highly desirable to attempt to do so in view of the chemical significance of the method. Some molecules present no problems, e.g. the case of ethylene is straightforward. There are only two π electrons and the excited state can be described in terms of ionic structures which are normally omitted. It is described by a function containing equal weights of the two possible ionic structures:

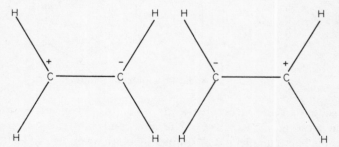

FIGURE 30. Ionic structure for ethylene.

When we go to more complicated π electron systems the problem is no longer straightforward. For butadiene it is necessary to use structures which correspond to the higher energy states which arise out of the ground state problem together with ionic states such as $C^+ - C = C - C^-$ and $C^- - C = C - C^+$. As we go to more complicated molecules the task becomes much more difficult.

One very important application of this approach is in the understanding of the behaviour of molecules containing fixed double bonds upon excitation. The $C = O$ bond in its ground state is probably heavily weighted with the structure $C^+ - O^-$, whilst in the excited state it probably contains a large weighting of $C^- - O^+$. Where this type of ionic structure predominates in the excited state we speak of an "intramolecular charge transfer mechanism upon excitation". In this case, however, the structure of the excited state probably goes no further than correcting the polar nature of the group in the ground state, giving a structure which is to all intents and purposes covalent.

In describing the ground states of conjugated and aromatic molecules only covalent structures have been used. When we seek to describe the lower excited states of these molecules, do we use the higher energy states which emerge in the ground-state calculation or do we admit ionic structures into the excited state wave function? Most of the calculations which have been carried through on this class of molecule have used only purely covalent structures.

In the table below are shown calculated values of the first electronic excited state for the polyene series; in which all covalent structures have been included.

	Calculated (α)	Observed (cm^{-1})
Butadiene	3·46	46,080
Hexatriene	2·61	39,750
Octatetraene	2·07	33,300

These data yield a value of $15,000 \ cm^{-1}$ or $43 \ kcal \ mole^{-1}$ for α. This is about three times the value required to fit the resonance energy data.

For the polyacene series we obtain (again using only covalent structures).

	Calculated (α)	Observed (cm^{-1})
Benzene	2·40	37,900
Naphthalene	1·97	31,800
Anthracene	1·60	26,400
Naphthacene	1·31	21,200
Pentacene	1·08	17,400

This series of data yield a consistent value of 16,200 cm^{-1} for α. Further it should be noted that this is in excellent agreement with the value found for the polyenes.

5.17 ASSESSMENT OF VALENCE BOND THEORY

In comparing the valence bond theory with the molecular orbital theory certain basic facts must be borne in mind. First, the molecular orbital theory is wont to overemphasize the significance of ionic structures in the wave function, whilst the valence bond approach underestimates the contribution of ionic structures to the wave function. Secondly, although the valence bond theory is conceptually attractive to minds trained to think in classical chemical terms, it loses its appeal as soon as it is necessary to include extensive ionic structures in the wave function. Indeed it has been shown recently that it is the presence of these ionic structures which leads to the stability of molecules which are conceived in a classical chemical picture as purely covalent. On the other hand, the molecular orbital theory is often difficult to interpret in terms of what is going on when reacting species meet under conditions in which reaction will occur. Thirdly, it is easier to carry through elaborate calculations using the molecular orbital theory whilst these become very tedious in the valence bond theory.

In short, it should be noted that both theories fail to predict electronic properties in a quantitative way when used in their least sophisticated forms. Molecular orbital theory works fairly well in its most sophisticated form but it is almost impossible to carry through valence bond theory to the same level of sophistication. Both theories, in their simplest form, enable a qualitative understanding to be had of what is going on.

5.18 FREE-ELECTRON THEORY

There is one other approach to understanding the behaviour of π electrons which is of great conceptual importance. This is the Free-Electron Theory. In this theory the π electrons are free to move over the whole of the nuclear framework which thus constitutes a "box" in which the electrons move. The theory is then exactly that of the "particle in a box" problem which was discussed in Chapter 1. It is assumed that the potential energy is constant over the whole range of the molecular framework, and the electrons are confined to this region by a barrier of infinite potential energy.

In a polyene molecule the electron is free to move over the whole length of the molecule. The box in which the electron is constrained must be longer than the molecule, because if this were not so the wave function at the end of the box (which would coincide with the terminal atoms) would vanish, there then being no probability of finding an electron in the region of the terminal atom. This is clearly in contradiction to physical reality, and it has become conventional to place the edge of the box one bond length outside each terminal atom.

For the linear polyene the wave equation is

$$-\frac{h^2}{8\pi^2 m}\frac{\mathrm{d}^2\psi}{\mathrm{d}x^2} = E\psi \qquad (0 \leqslant x \leqslant a) \tag{46}$$

where a is the length of the box and x is a point inside of it. The solution of this equation is

$$\psi_a = \sqrt{\frac{2}{a}}\sin\frac{n\pi}{a}x \qquad (0 \leqslant x \leqslant a) \tag{47}$$

n is a quantum number and can take any integral positive value. For each value of n, the corresponding energy is

$$E_n = \frac{n^2 h^2}{8ma^2} \tag{48}$$

In a conjugated molecule containing k double bonds there are $2k$ π electrons. These are just sufficient to fill k levels in the ground state. The first excited state arises when one of these electrons is promoted from the kth to the $(k+1)$th level. The transition energy is

$$\Delta E = E_{k+1} - E_k = \frac{(k+1)^2 h^2}{8ma^2} - \frac{k^2 h^2}{8ma^2}$$

$$= \frac{(2k+1)h^2}{8ma^2} \tag{49}$$

which in terms of wave numbers is

$$\Delta E = \frac{(2k+1)}{33a^2} \text{ cm}^{-1} \tag{50}$$

where a is expressed in Ångstroms. Setting the average polyene bond length to 1·4 Å, (50) becomes

$$\Delta E = \frac{10^7}{64·68(2k+1)} \text{ cm}^{-1} \tag{51}$$

The energy of the lowest band for the first four polyenes is given below

	Predicted energy (cm^{-1})	Observed energy (cm^{-1})
Ethylene	51,500	61,500
Butadiene	31,000	46,100
Hexatriene	22,100	39,750
Octatetraene	17,250	35,000

It is clear that the free-electron method does not give good predictions of polyene spectra. As a qualitative guide it is fairly satisfactory, and conceptually it gives mathematical expression to the Hückel idea of a sea of π electrons moving over the σ-bonded framework. It is unsatisfactory because it predicts a limit of zero for the first band as the chain becomes infinite.

The method can be applied to polycyclic aromatic molecules, by distorting the molecular framework so that the atoms lie on the circumference of a circle. Under these conditions problems of determining the box edge are removed, the only condition now being that of continuity of the wave function around the circumference of the circle. The wave functions for this type of "box" are easily obtained and are

$$\psi = \exp(im_q\phi) \tag{52}$$

where $m_q = 0, \pm 1, \pm 2$, etc., and ϕ is the polar angle about the centre of the circle. The energies of these wave functions are given by

$$E = \frac{q^2 h^2}{2ml^2} = 1·2 \times 10^6 q^2/l^2 \quad \text{cm}^{-1} \tag{53}$$

where l is the circumference of the perimeter in Ångstroms and q is a quantum number having all positive integral values, and is a measure of angular momentum about the central axis of the electron. The energy varies with q^2 and all levels are doubly degenerate (except $q = 0$), depending upon whether the electron moves in a positive or negative sense. In a molecule containing $4n + 2$ electrons (e.g. the polyacenes) the z component of the angular momentum will be zero.

The selection rule for transitions is $\Delta m_q = 1$. Transitions of the type $m_q \rightarrow m_q + 1$ and $-m_q \rightarrow -(m_q + 1)$ are allowed, whilst transitions of the type $m_q \rightarrow -(m_q + 1)$ are forbidden.

Agreement of this type of calculation with experiment is reasonable. For the first four bands of naphthalene which have band origins at 31,800, 34,600, 45,500 and 52,600 cm^{-1}, the free-electron model predicts 36,600, 41,000, 48,900 and 51,200 cm^{-1} respectively.

It is not easy to understand other molecular properties in terms of this model. Nevertheless the model describes in simple mathematical terms the physical reality of the delocalized electrons moving over the σ-bonded framework. The model fails because it does not take into account the periodicity in the potential energy due to the ionic framework.

5.19   CONCLUSION

In this chapter we have examined in some considerable detail the application of both the molecular orbital and valence bond approach to the understanding of the π-electron properties of conjugated, aromatic and heterocyclic molecules, and also the free-electron mode. It has been seen that all three methods give a good qualitative interpretation of the properties of these molecules.

The free-electron model does not lend itself to any further sophistication. On the other hand the valence bond method can be improved both in its formulation of the π-electron interactions and in the number of structures included. The resultant problem becomes intractable as far as computation is concerned, and, in achieving greater degrees of sophistication, the essentially straightforward picture is lost.

The theory which lends itself most easily to further sophistication without loss of ease of interpretation is the molecular orbital interpretation. It is possible to develop a molecular orbital theory in which π-electron interactions are properly included in the Hamiltonian, and still retain the essential features of the Hückel model. Some recent work suggests that it may be possible to formulate a theory to deal with the σ electrons, which is similar in approach to that developed for the π electrons. This is, however, still in its early stages of development.

In this chapter we have discussed the details of calculations in a full manner because they are so simple to carry through, and also because of the great general interest in the electronic structure of organic molecules.

6
Inorganic Molecules

6.1 INTRODUCTION

There is still a wide variety of molecules we have not yet referred to whose structures can be discussed in terms of the ideas used in previous chapters. There is, for example, a wide range of molecules where there would appear to be insufficient electrons to form stable bonds, and this occurs in many of the boron hydride molecules. There are inorganic molecules which have sufficient electrons to form stable bonds yet nevertheless have incomplete valence shells.

It is possible to have molecules with aromatic character which do not contain carbon. These molecules have ring structures made up of either boron and nitrogen atoms or else of phosphorus and nitrogen atoms. This extension of aromaticity to molecules which contain no carbon atoms has led to a deepening of our ideas about aromaticity.

There are also many inorganic molecules known as complexes, which involve transition element atoms or ions surrounded frequently by six ligand ions or highly polar groups. It used to be thought that the structure of such a molecule could be understood either in terms of highly ionic species or else by using extremely involved covalent bonding schemes. In recent years there has been considerable interest in this type of molecule, and much contemporary inorganic chemistry is concerned with this field.

This chapter will examine briefly some of the ideas on the bonding of molecules which fall into these three types.

6.2 ELECTRON-DEFICIENT MOLECULES

Consider the molecule BH_3. The electronic configuration of the boron atom is $(1s)^2(2s)^2(2p)$. Without hybridization it would be expected that boron would be univalent forming one bond (with, say, hydrogen to give

BH). This molecule would not be very stable because the $p\sigma$ bond is not very strong. In fact, except in unstable spectroscopic fragments, such molecules are unknown. If, however, one electron from the $2s$ orbital is excited into an empty $2p$ orbital, a new possibility occurs. The boron atom is trivalent, forming one $s\sigma$ bond and two $p\sigma$ bonds. In the BX_3 molecules which are known the three B—X bonds are equivalent. The boron must therefore be in a valence state of sp^2 hybridization. This leads to three σ bonds, equivalent with each other, planar and strong. This is in accord with the experimental evidence. On the boron atom this leaves an empty p orbital perpendicular to the plane of the molecule. In other words, the BH_3 molecule would be electron-deficient, the boron atom being unable to complete a closed shell configuration.

The trihydride BH_3 does not exist and the reason will become apparent after an examination of the structure of BF_3 which does exist. The electronic configuration of the fluorine atom is

$$(1s)^2(2s)^2(2p)^5$$

At first sight it appears that the bonding is due to a p electron from the fluorine atom. However sp mixing occurs to allow strong σ bonding between the fluorine atom and the central boron atom. This leaves a diagonal lone pair and two p lone pairs perpendicular to the B—F axis. One of these lone pairs on each fluorine atom can align itself so as to form a π-electron system built out of one p orbital on each fluorine atom and the vacant p orbital on the boron atom. There are available six p electrons for this system. Out of the four p orbitals, four π molecular orbitals can be constructed. Three of these will be doubly occupied. The vacant boron p orbital will thus have its electron deficiency remedied to some extent by a partial "sharing" in the lone pairs on the fluorine atoms. This explanation accounts for the structure of all of the boron halides.

In the boron trihydride molecule BH_3 this π bonding is impossible, for the hydrogen atoms are unable to contribute to the π-electron system. The BH_3 molecule does not therefore exist. Nevertheless BH_3CO does exist. Consider the structure of carbon monoxide. The carbon atom is in a state of digonal hybridization. It uses one of its electrons to form a σ bond with the oxygen atom. One of the remaining electrons forms a π bond with an electron in a half-filled orbital on the oxygen. A second π bond is formed by the overlap of one of the oxygen lone pairs with a vacant p orbital on the carbon atom. This leaves two electrons on the carbon atom in the other digonal hybrid. In the carbon monoxide molecule this digonal orbital containing two electrons constitutes a lone pair. It is this lone pair which is able to enter into bonding with the BH_3 fragment.

The boron atom becomes tetrahedral, three of the four sp^3 hybrids, containing one electron each, form the three B—H bonds whilst the fourth overlaps with the lone pair on the carbon atom to give a strong σ bond. This sort of bond is of course highly polar. In this way the boron atom is able to attain a closed shell configuration.

This kind of bonding, in which both electrons are provided by the lone pair of electrons on the carbonyl carbon atom, is able to account for the structure of many of the carbonyl complexes with metal atoms or ions.

Although BH_3 does not exist, many boron hydride molecules do. The simplest, and in some ways, the most interesting, is B_2H_6. For many years it was assumed that this molecule has the same structure as ethane. This is of course impossible because the two molecules are not isoelectronic. Various attempts were made to account for the deficiency of electrons. The existence of one-electron bonds was postulated, whilst others suggested that one bond is completely missing, although this must be averaged out, giving each bond six-sevenths the strength of a normal single bond. When the spectrum of B_2H_6 was examined it was found that the molecule does not have the three-fold rotation axis of symmetry which is possessed by ethane. It possesses only a two-fold axis. This requirement lead to a rethinking of the possibilities for the structure of this molecule. Coupled with the fact that chemical evidence suggests that only four of the hydrogen atoms are equivalent, the tentative structure

was suggested, in which the four extreme hydrogen atoms together with the boron atoms are coplanar whilst the other two hydrogen atoms are perpendicular to the plane of the molecule. This structure requires that the boron atoms be in an sp^2 state, the H—B—H bond angle being 120°. Measurements of the in-plane bond angle do in fact agree with this, for they are found to be 120°. The other H—B—H bond angle for the out-of-plane system is found to be 100° and the interatomic B—B distance to be not all that much greater than one would expect for a B—B double bond analogous to that in ethylene. How can this be when there are insufficient electrons to achieve it?

Many solutions have been suggested to this problem, using both molecular orbital and valence bond theory, and it is instructive to examine some of them.

Within the framework of the valence bond theory there are two possibilities for the structure. It can be explained in purely covalent terms using the structures

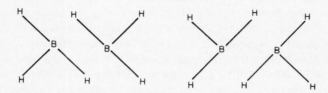

This explanation is not satisfactory, for the resulting resonance energy is insufficient to compensate for the distortion energy needed to realize the valence angles.

Another possibility is that considerable ionic weight must be given to the valence bond wave function, and that this can arise through the participation of structures of the type

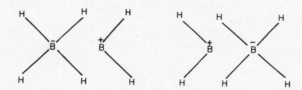

Once again it is difficult to see how the necessary stability can be achieved in the presence of a geometrical framework which requires such considerable distortion.

One possible explanation of the structure is to say that the two BH_2 groups are joined together by the two "bridge" hydrogen atoms through electrostatic "hydrogen" bonding, as is known to exist in solution in many cases. This is unlikely because of the energy required to make the entity stable and also because the electronegativity of the hydrogen atom is not significantly different from that of boron.

An ingenious suggestion came from K. S. Pitzer, who considered the molecule analogous to ethylene. The remaining two hydrogen atoms were considered to give up their electrons to form the π bond between the two boron atoms. The two protons were then retained in the structure by being held in the cloud of electrons set up in the π bond. This theory is very attractive because it does explain the fact that two of the electrons lie between the boron atoms yet above and below it. It does account for the fact that the interatomic distance between the boron atoms is 1·77 Å which is only about 0·15 Å greater than the expected distance for a double bond. There is of course sufficient overlap between the p orbitals to give a stable π bond.

Against this structure is the fact that no other case is known. Such structures have been proposed as intermediates in addition reactions involving ethylene, but none of these have ever been isolated or detected spectroscopically. Secondly it would be expected that these protons would be extremely labile, whereas none of the hydrogen atoms in this molecule are labile. A third objection arises from a consideration of the reactivity of diborane. Study of the reactions of diborane suggest that the reactive species is H^- rather than H^+. This evidence requires us to set aside this explanation of the structure in spite of its many attractive features.

The solution to the problem was seen by H. C. Longuet-Higgins who was able to show that Pitzer's wave function was amenable to another interpretation. Instead of considering the structure as basically ethylenic, the wave function can be reinterpreted in terms of two BH_2 units and two other orbitals which are tricentric, each containing two electrons. This structure corresponds to

in which each boron is engaged in four bonds, two straightforward σ bonds with the in-plane hydrogen atoms, and two bonds perpendicular to these in which the bond angles are 100°, covering both boron atoms and one hydrogen atom. Both of these "sausage" bonds are equivalent. This interpretation of Pitzer's wave function does away with the necessity for isolated protons in the structure. It involves the hybridization being approximately tetrahedral with distortions to take into account the difference in electrostatic repulsions between non-equivalent pairs. It is probable that this is the most satisfactory explanation put forward for the structure of this fascinating molecule.

The only other plausible explanations involve the participation of d orbitals but these can be discounted, for d orbital participation is of no significance in such light atoms.

6.3 INORGANIC AROMATICS

Hückel showed that a planar monocyclic molecule is aromatic in character if it contains $(4m + 2)$ π electrons where m is any integer. When m is one, a molecule with six electrons is aromatic. The table below shows the number of π electrons which would convey aromatic character on a monocyclic system.

m	1	2	3	4
n	6	10	14	18

The obvious example of a molecule with six π electrons is benzene. The rule does not say, however, that all of the atoms in the ring must be identical, only that they must use the same sort of orbital in the construction of the molecular orbitals. In the organic field a molecule which is aromatic but which does not have only one kind of atom is *sym*-triazine.

This molecule has aromatic character and contains alternate carbon and nitrogen atoms, with each atom donating one electron to the π-electron system.

In its formulation, Hückel's law requires the existence of degenerate molecular orbitals, and this requires an axis of at least three-fold rotational symmetry. This excludes many well-known molecules with aromatic character, including pyrrole

In this case the nitrogen atom donates two electrons to the system and thus completes the "aromatic sextet".

On the other hand Hückel's rule states that monocyclic molecules with only $4m$ π electrons are not aromatic. This excludes molecules like cyclobutadiene

which has no delocalization energy (as a simple calculation confirms) and cyclooctatetraene

Many of the important aromatic molecules have the general basic formula $(AB)_n$ where A and B are different atoms. Such systems have $2n$ orbitals taking part in the π-electron system. With second row elements like boron, carbon and nitrogen, only p orbitals are capable of forming π orbitals. If we go to the next row, however, and take phosphorus or sulphur atoms it is then possible to use d orbitals in the construction of π orbitals.

The classic example of an aromatic inorganic molecule is borazole, which is formed by heating together boron trichloride and ammonium chloride and reducing the intermediary. Borazole has the formula $B_3N_3H_6$ and it has the structure

The nitrogen atoms each contribute two electrons to the π-electron system whilst the boron atoms have a completely empty p orbital. There are thus six π electrons and the molecule is aromatic. This is an example where the orbitals are all of $2p$ character.

Systems in which p and d orbitals both contribute to the aromatic system are provided by the phosphonitrilic halides $(NPX_2)_n$ and the thiazyl halides $(NSX)_n$, the simplest examples being

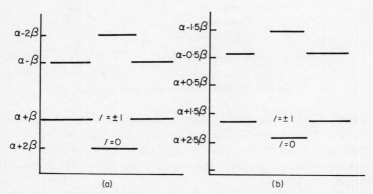

In the case where the orbitals are all p orbitals, like benzene, *sym*-triazine or borazole, the molecular orbitals occur in pairs, with energies which are the roots of the secular equation

$$\begin{vmatrix} (\alpha_A - E) & 2\beta \cos\dfrac{l\pi}{n} \\ 2\beta \cos\dfrac{l\pi}{n} & (\alpha_B - E) \end{vmatrix} = 0$$

where l is an integer having the value $0, \pm 1, \ldots, n-1/2$; and α_A and α_B are the Coulomb integrals appropriate to the $2p$ orbitals of atoms A and B respectively. β is the resonance integral appropriate to the A—B bond. The orbital energy level sequence for benzene and for borazole (where $\alpha_N = \alpha_B + \beta$) are shown below in figure 31.

FIGURE 31. Orbital energy sequence for (a) benzene, (b) borazole.

In the case of A being a d orbital and B a p orbital a significant difference arises. This can be best understood in a pictorial way. Consider a p orbital with d orbitals from A neighbours on each side. In terms of the signs of the wave functions we have the following situation

$$
\begin{array}{ccccc}
+ & - & + & + & - \\
\cdot & & \cdot & & \cdot \\
- & + & - & - & + \\
A & & B & & A \\
d_1 & & p & & d_2
\end{array}
$$

This gives rise to resonance and overlap integrals $S_{d_1 p}$ and $S_{p d_2}$ which have opposite signs. This leads to a secular determinant of the form

$$
\begin{vmatrix}
\alpha_A - E & 2i\beta \cos \dfrac{l\pi}{n} \\[3mm]
-2i\beta \cos \dfrac{l\pi}{n} & \alpha_A + \beta\delta - E
\end{vmatrix} = 0
$$

Whilst the energy levels still occur as degenerate pairs, both the position and order are changed compared with the p–p case. The sequence of energy levels is shown in figure 32. On the left we have plotted the

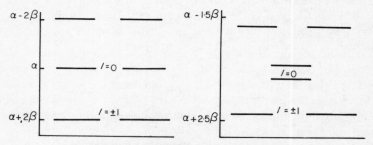

FIGURE 32. Energy levels for $p\pi$-$d\pi$ systems.

physically unreal case of equal electronegativity ($\alpha_A = \alpha_B$) in order to facilitate comparison with the p–p case; on the right is plotted the case for $\alpha_B = \alpha_A + \beta$. If we had chosen eight-membered rings instead of six-membered rings, then merely inversion of the energy levels would have occurred.

Studies on inorganic aromaticity are still in the early stages but it is still probably true to say that a whole new field has opened up for investigation by both experimenters and theoreticians.

With the introduction of d orbitals many new possibilities present themselves, and there is much greater scope for constructing hybrid orbitals and bonding possibilities.

6.4 TRANSITION METAL COMPLEXES

Molecular entities involving a transition metal ion "bonded" to four or six ions or polar ligands present many problems in understanding their electronic structures. In the 1930's attempts were made to understand the structure of these complexes in terms of valence bond theory and strongly-directional hybrid orbitals having considerable d orbital participation. This work is associated with Linus Pauling. In the 1950's there emerged a re-examination of some work done in the early 1930's on the effect of a crystal field with octahedral or tetrahedral symmetry on a transition metal ion. This approach considered the ligand atoms as point charges. Use of this theory and its modified forms has led to considerable progress in the understanding of the structure of these complexes. Simultaneous with this work another line of attack was developed, which was to try to understand the structure of complexes in terms of molecular orbital theory.

6.4a The valence bond method

The common feature which underlies all of this category of molecules is the presence of unfilled shells of electrons in d orbitals. Mixtures of s, p and d orbitals which will give hybrids having σ character and directed towards the corners of a regular octahedron can be built out of $3s$, $3p_x$, $3p_y$, $3p_z$ and two d orbitals, the $d_{x^2-y^2}$ and d_{z^2}. The d_{z^2} orbital is strongly directed along the z axis and the $d_{x^2-y^2}$ orbital is in the xy plane and has its lobes directed along the x and y axes. The sp^3d^2 hybrid orbitals are six orbitals built out of these atomic orbitals and directed towards the corners of a regular octahedron.

Consider the cobalt ion Co^{3+}, which has six electrons in the $3d$ subshell with the $4s$ and $4p$ subshells vacant. The free ion has two electrons in one d orbital and one in each of the other four $3d$ orbitals. In order to leave the two d orbitals which are necessary for the hybridization process, it is necessary to place the two electrons in the necessary orbitals into the other two half-filled d orbitals. This then leaves the two $3d$, the three $4p$ and the $4s$ orbitals vacant. In this state all of the cobalt electrons are paired and the ion will not be paramagnetic, having no unpaired electrons. Six NH_3 molecules can then attach themselves to the cobalt ion through forming bonds involving the lone pair of electrons on the nitrogen atom and one of the empty hybrids on the cobalt ion. There are thus six bonds formed between the ammonia molecules and the cobalt ion; each of these is highly polar.

In the case of chromium complexes involving Cr^{3+}, in the valence shell there are three electrons which half-fill three of the $3d$ orbitals. This leaves the remaining two d orbitals together with the $4s$ and three $4p$ orbitals for the formation of the octahedral hybrids. Ammonia molecules are then able to complex with the chromium ion filling the vacant hybrid orbitals with electrons donated by the ligand molecules. This leaves three half-filled d orbitals on the central ion, and the $[Cr(NH_3)_6]^{3+}$ ion is paramagnetic with a susceptibility in accordance with the existence of three unpaired electrons.

Whilst there are many attractive features to this theory, it accounts for little else beyond the paramagnetism of the complexes.

6.4b   The crystal field method

In order to understand the development of this theory, it is necessary to have some conception of the directional properties of the five d orbitals in a free atom. The directional properties of the wave functions corresponding to the d orbitals are shown in figure 33.

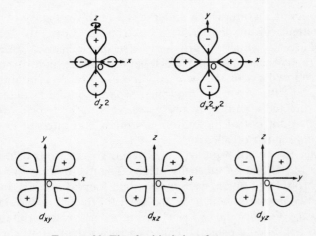

FIGURE 33. The d orbitals in a free atom.

Simple examination shows that the three orbitals xy, xz and yz have similar direction properties and can be converted into each other by simple rotation of the axes. The z^2 orbital is different in as much as the bulk of the electron density in this orbital is directed along the z axis. The remaining orbital can be considered to be constructed out of x^2 and

y^2 orbitals, the positive combination being contained in the other three orbitals.

The essence of the crystal field method is to examine the effect of a field of high symmetry on the five-fold degeneracy of the d orbitals which exists in the free atom.

Consider the effect on the d orbitals of an atom M of six point charges placed about M, in such a way that the point negative charges are at the corners of a regular octahedron. Let these negative charges lie on the x, y and z axes as shown below in figure 34.

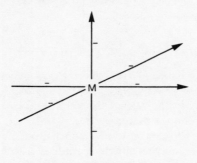

FIGURE 34. An octahedral field.

Consider first the effect of the octahedral field on the z^2 orbital. The large lobes are directed along the axis leading to considerable repulsion between the orbital and the point charges on the z axis. Similarly with the x^2-y^2 orbital, since the direction of maximum density lies along the x and y axes there will be repulsion between the point charge and the electron in the orbital. In these cases the orbital energy will be raised. In fact they are raised by the same amount in each case leading to a degenerate pair, usually given the symbol e_g.

The other three d orbitals xy, xz and yz all have their directions of maximum density lying between the axes. All three cases are identical except for simple rotation in space. This will cause attraction between the now only partially-screened nucleus M and the point charges, and thus lead to a depression of the energy of the d orbital. Since all three orbitals are affected in the same way, this will lead to a triply degenerate energy level, usually given the symbol t_{2g}. As with configuration interaction, the sum of the energies of the new levels is the same as the sum of the energies of the five d orbitals in the field-free case. Suppose the energy of the d orbital in the field-free case is E_0, and the energy of the e_g pair is $E_0 + x$, then the energy of the t_{2g} orbitals is $E_0 - \frac{2}{3}x$. This then makes the sum of the five orbitals $5E_0$. This sort of splitting in which the "centre of gravity" of the set of levels is preserved in general for any splitting in which the forces are purely electrostatic, and where the set of energy levels considered is well removed from any other set of energy levels. The splitting pattern for the five energy levels in the octahedral field is shown in figure 35.

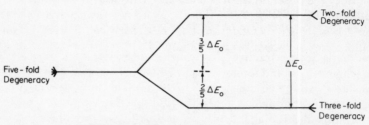

FIGURE 35. Splitting of d orbital energies in an octahedral field.

This explains why in the case of the Co^{3+} ion, two d orbitals are completely filled and the ion is diamagnetic.

The case in which the central ion M is surrounded by four point charges at the vertices of a regular tetrahedron leads to a totally different pattern of splitting. In this case the point negative charges lie midway between the axes, if we choose our axes in the usual way to pass through the centres of the faces. This leads to the raising of the energy of the three orbitals xy, xz and yz and to a depression of the energy of the pair of orbitals z^2 and x^2-y^2. The energy levels will be inverted with respect to the octahedral case. The splitting for the tetrahedral case is shown in figure 36.

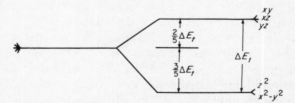

FIGURE 36. Splitting of d orbitals in tetrahedral field.

If cation–anion distances are the same in both the octahedral and tetrahedral cases, it is easy to see that the splitting of the levels in the tetrahedral case is about half that of the splitting in the octahedral case. It is in fact $\Delta E_t = \frac{4}{9} \Delta E_0$.

The picture which arises in the crystal field case allows us to understand the electronic spectra of these complexes. In the case of an atom or ion which has either one or nine d electrons, the transition energy will be the splitting between the two sets of energy levels. The spectra of systems in which there are two to eight d electrons are much more complicated, as the energies of the various possible spectroscopic states are functions of the splitting ΔE. Further complications arise in these intermediate cases because of degenerate ground or excited states leading to a Jahn–Teller effect (see Chapter 4, section 7) and the geometry of the complex changing slightly to remove the degeneracy.

This way of considering the structure of inorganic transition metal complexes is very attractive. Nevertheless there are certain objections to the scheme. First and foremost is the fact that real ligands are not point charges and cannot be considered to be so. The only case which can justify such an approach is that when the overlap between the central atom orbitals and the ligand orbitals is small. In many cases this is so, and to a fair degree of accuracy the crystal field approximation holds. There are cases, however, where the overlap is not small, and in these cases other methods of approach must be used. This is just another way of saying that to a greater or lesser extent a measure of covalent bonding occurs in these complexes.

6.4c The molecular orbital method

The valence bond explanation of the structure of transition metal complexes has already been discussed. It is now necessary to turn our attention to the molecular orbital approach to bonding in these complexes.

There are nine orbitals available for bonding, in the case of the first set of transition metal elements. These are the $4s$, three $4p$ and the five $3d$ orbitals. Of these nine orbitals, six are capable of forming σ bonds. These are the $4s$, the three $4p$, and the $3d_{z^2}$ and $3d_{x^2-y^2}$. These orbitals are capable of participating in σ bonding because their lobes lie along the axes, and these are the best directions for σ bonding in the octahedral case. The remaining three orbitals xy, xz and yz all lie skew to the internuclear axes. These atomic orbitals are able to form π orbitals with p (or d) orbitals on the ligand atom or groups. In the molecular orbital approach we see the possibility for multiple bonding in these complexes.

In the case of tetrahedral complexes a different set of possibilities arise for π bonding.

The molecular orbital method is thus able to account for the bonding in a more general way than was possible in the valence bond approach, but its shortcomings arise when we attempt to calculate the energy levels of the orbitals. On the other hand it gets over the problem of regarding the ligands as point charges as is assumed in the straightforward crystal field method.

The molecular orbital approach is capable of interpreting the electronic properties of the complex. In particular the electronic spectra of the complexes can be interpreted in the usual way.

It is necessary to say a little more at this stage about the extent of the splitting which is predicted in the crystal field theory. In general this lies in the range 10,000 to 30,000 cm^{-1}, and this places spectral transitions in the region of 3000 to 10,000 Å. In other words most of these transitions lie in the visible spectrum, and this is the reason why many of these complexes are coloured.

There are still many unsolved problems in the understanding of these inorganic complexes, and most of these are associated with the great difficulty of obtaining reliable wave functions.

6.5 CONCLUSION

In this chapter we have discussed, albeit in a rather cursory way, some of the problems of interest to the theoretician in modern inorganic chemistry. The problems connected with electron-deficient molecules such as the boron hydrides raises fascinating theoretical problems which after many years effort are still unsolved. On the other hand the recognition of aromaticity in inorganic molecules such as the borazoles and phosphonitriles have opened a complete new vista in theoretical chemistry. In the third field which we examined, that of the transition metal complexes, there arise problems which seem almost incapable of solution.

There was a time when inorganic chemistry was considered to be a closed book. It is inconceivable that that particular time was only about twenty years ago. In the time which has elapsed since then all of the fields mentioned above have opened up. It would seem that little progress will be made without elaborate computation, but if this field is to have a more solid theoretical foundation much fundamental thinking needs to be done.

One field of interest which has not been mentioned in this chapter is that of organometallic compounds, where the metallic atom is not necessarily a transition metal atom or ion. This is a field where inorganic and organic electronic theories must be merged. It could be that in tackling problems of this kind, new methods will emerge of considerable importance in quantum chemistry.

Bibliography

General texts on quantum mechanics

The two best elementary texts are

Quantum Chemistry, by H. Eyring, J. Walter and G. E. Kimball, John Wiley and
Sons, New York, 1944.

Introduction to Quantum Mechanics by L. Pauling and E. B. Wilson Jnr., McGraw-
Hill, New York, 1935.

For more advanced students who have considerable mathematical ability there are

Quantum Mechanics, Vols. 1 and 2 by A. Messiah, North Holland Publishing Co.,
Amsterdam, 1962.

Quantum Mechanics by P. A. M. Dirac, Oxford University Press, 4th Edition, London,
1958.

Atomic structure and spectra

Quantum Theory of Atomic Structure, Vols. 1 and 2, by J. C. Slater, McGraw-Hill,
New York, 1960.

These books are of great significance, because in addition to their worth as
textbooks, they contain extensive bibliographies.

Atomic Structure and Atomic Spectra, by G. Herzberg, Dover Publications, New
York, 1950.

Small molecules

By far the best introduction to valence theory is *Valence*, by C. A. Coulson,
Oxford University Press, 2nd Edition, London, 1961.

Of considerable historic interest, although extremely biased in favour of the
Valence Bond Theory is

The Nature of the Chemical Bond, by L. Pauling, Cornell University Press, 2nd
Edition, New York, 1960.

At a more advanced level, and containing an extensive bibliography there is

Quantum Theory of Molecules and Solids, Vol. 1, by J. C. Slater. McGraw-Hill,
New York, 1963.

There is now a vast collection of books on molecular spectra. The best introduction
is

Spectroscopy and Structure, by R. N. Dixon, Methuen, London, 1965.

The standard work on molecular spectroscopy is
Infra-Red and Raman Spectra, 1945.
Spectra of Diatomic Molecules, 1950.
Electronic Spectra of Polyatomic Molecules, 1966, all three volumes by G. Herzberg,
    Van Nostrand, New York.

Aromatic and heterocyclic molecules

There are now many texts on this subject. Among those available the student may
care to consult
The Molecular Orbital Theory of Conjugated Systems, by L. Salem, Benjamin, New
    York, 1966.
The Electronic Properties of Aromatic and Heterocyclic Molecules, by T. E. Peacock,
    Academic Press, London, 1965.
Molecular Orbital Theory for Organic Chemists, by A. Streitweiser, John Wiley and
    Sons, New York, 1961.
Concerned with the spectra of organic molecules, we have
Theory and Applications of Ultraviolet Spectroscopy, by H. H. Jaffe and M. Orchin,
    John Wiley and Sons, New York, 1962.
Theory of the Electronic Spectra of Organic Molecules, by J. N. Murrell, Methuen,
    London, 1963.

Inorganic molecules

For a discussion on aromatic character and its extension to inorganic molecules
see the article entitled
Aromatic Character, by D. P. Craig, in *Theoretical Organic Chemistry*, Butterworths
    Scientific Publications, London, 1959.
For a general introduction to the theory of transition metal complexes see
An Introduction to Transition Metal Chemistry, by L. E. Orgel, Methuen, London,
    1960.
Orbitals in Atoms and Molecules, by C. K. Jorgenson, Academic Press, London,
    1962.
Advanced Inorganic Chemistry, by F. A. Cotton and G. Wilkinson, Interscience,
    2nd Edition, London, 1966 (especially Chapter 26).

Index

KRAUSKOPF LIBRARY

541.28 P313   STACKS
Peacock, T. E./Foundations of quantum ch

3 1896 00008 1640